New Views on R. Buckminster Fuller

New Views on R. Buckminster Fuller

Edited by Hsiao-Yun Chu and Roberto G. Trujillo

Stanford University Press
Stanford, California

Stanford University Press
Stanford, California

This book has been published with the assistance of Stanford University Libraries.

Printed in the United States of America on acid-free, archival-quality paper

Library of Congress Cataloging-in-Publication Data

New views on R. Buckminster Fuller / edited by Hsiao-Yun Chu and Roberto G. Trujillo.
p. cm.
Includes bibliographical references and index.
ISBN 978-0-8047-5209-1 (cloth : alk. paper)--ISBN 978-0-8047-6279-3 (pbk. : alk. paper)
1. Fuller, R. Buckminster (Richard Buckminster), 1895-1983--Influence. 2. Engineers--United States. 3. Architects--United States. 4. Inventors--United States. I. Chu, Hsiao-yun. II. Trujillo, Roberto G., 1951-
TA140.F9N48 2009
620.0092--dc22 2009007192

Designed by Bruce Lundquist
Typeset at Stanford University Press in 10/15 Minion

Contents

Contributors

Hsiao-Yun Chu is assistant professor in the Department of Design and Industry at San Francisco State University, where she teaches product design and the history of design and technology. Prior to this position, she spent two years as the assistant curator of the R. Buckminster Fuller Collection at Stanford University, where she helped to build an online digital media collection to complement the physical archive and began research for this book. She holds degrees from Harvard and Stanford universities.

Maria Gough is associate professor of Modern and Contemporary Art at Stanford University. A specialist of the Russian and Soviet avant-garde, Gough has published her research in numerous journals and exhibition catalogs, and her book *The Artist as Producer: Russian Constructivism in Revolution* was published by the University of California Press in 2005. Her current research concerns the work of El Lissitzky and Gustavs Klucis, extreme drawing practices in the twentieth century, and a study of the "afterlife" of the 1920s in the 1950s and 1960s.

Barry M. Katz teaches in the Department of Industrial Design and the Graduate Program in Design at California College of the Arts. He has been an editor at *Design Book Review* and a contributing editor for *I.D.* magazine and *Metropolis.* Barry's books include *Herbert Marcuse and the Art of Liberation* (New York: Schocken Books, 1982); *Foreign Intelligence: Research and Analysis in the Office of Strategic Services, 1942–1945* (Cambridge, MA: Harvard University Press, 1989); and *Technology and Culture: A Historical Romance* (Stanford, CA: Stanford University Press, 1990). His new book, in progress, examines the history of Silicon Valley design.

Joachim Krausse is professor of design theory in the Department of Design at Anhalt University of Applied Sciences in Dessau, where he heads the Master course in Integrated Design (MAID). His research on R. Buckminster Fuller, along with that of Claude Lichtenstein, led to a traveling exhibition (1999–2002) and book (in two volumes) entitled *Your Private Sky: R. Buckminster Fuller, the Art of Design Science* (Baden: Lars Müller, 1999). His research focuses on issues of technical culture and the importance of construction as an embodiment of intelligence.

Claude Lichtenstein is a writer and historian whose research on R. Buckminster Fuller, along with that of Joachim Krausse, led to a traveling exhibition (1999–2002) and a book entitled *Your Private Sky: R. Buckminster Fuller, the Art of Design Science* (Baden: Lars Müller, 1999). Additional books include *As Found: The Discovery of the Ordinary* (Baden: Lars Müller, 2001) and *Handmade* (Baden: Lars Müller, 2005).

Reinhold Martin is an architect and associate professor of architecture at Columbia University, where he directs the Master of Science Program in Advanced Architectural Design. He holds a PhD from Princeton University, as well as degrees from the Architectural Association and Rensslaer Polytechnic Institute. A founding coeditor of the journal *Grey Room*, he is the author of *The Organizational Complex: Architecture, Media, and Corporate Spaces* (Cambridge, MA: MIT Press, 2003).

Jonathan Massey is associate professor in the School of Architecture at Syracuse University, where he teaches the history of architecture. Prior to joining the Syracuse faculty, Massey taught at universities in Los Angeles and New York, and he worked at the architecture firms of Dagmar Richter and Frank O. Gehry & Associates. He holds bachelor and doctoral degrees from Princeton University and a Master of Architecture degree from UCLA. He is the author of a forthcoming book on Claude Bragdon's progressive architecture.

David E. Nye, PhD University of Minnesota, is professor of history at Warwick University. His ten books include *Electrifying America* (1990); *American Technological Sublime* (1994); *Consuming Power* (1998); *America as Second Creation* (2003); and *Technology Matters: Questions to Live With* (2006). In 2005 he received the Leonardo da Vinci Medal from the Society for the History of Technology, its highest honor.

Felicity D. Scott is assistant professor of architecture at Columbia University. She is a founding coeditor of *Grey Room*, a quarterly journal of architecture, art, media, and politics published by MIT Press since 2000. Her book *Architecture or Techno-Utopia: Politics after Modernism* was published by MIT Press in 2007.

Howard P. Segal is professor of history and director of the Technology and Society Project at the University of Maine. He specializes in the history of technology and the history of science, and his research and writing have focused on the history of American technology and of American utopianism. His publications include *Technological Utopianism in American Culture* (Chicago: University of Chicago Press, 1985) and *Recasting the Machine Age: Henry Ford's Village Industries* (Amherst: University of Massachusetts Press, 2005).

Fred Turner is an assistant professor and the director of undergraduate studies in the Department of Communication at Stanford University. He is the author of *Echoes of Combat: Trauma, Memory, and the Vietnam War* (Minneapolis: University of Minnesota Press, 2001) and *From Counterculture to Cyberculture: Stewart Brand, the Whole Earth Network, and the Rise of Digital Utopianism* (Chicago: University of Chicago Press, 2006). His research and teaching focus on digital media, journalism, and the intersection of media and American cultural history.

Preface

The R. Buckminster Fuller Papers were acquired by the Stanford University Libraries in 1999. This acquisition was a singular event for the Libraries, both because of the size and scope of the collection and because of the new critical readings of Fuller that we anticipated.

To be sure, the organization and description of the collection, and its continuing updates, were and are a significant challenge—and an important investment for intellectual access. True to original expectations, scholarly interest in Fuller's ideas and work and demand for access to the archive have continued unabated since 1999. The collection has inspired everything from ten-page undergraduate papers to major museum exhibitions, including the Whitney's *Buckminster Fuller: Starting with the Universe* (June 26–September 21, 2008, New York City). Dissertation-level research, undergraduate and graduate research seminars and classes, and scholarly articles—such as those included in this volume—have all found their beginnings in the Fuller collection. In keeping with Stanford's intention to provide for precisely such new critical readings of Fuller, we have sought to promote and encourage access in new and creative ways, including, for example, online access to full-length audio and video recordings from the archive. Stanford was not interested in simply acquiring a major collection and then hoping that someone would someday use it. The Libraries were fully expecting a new generation of scholarship around Fuller, and a spate of recent work on him has confirmed our hopes.

The volume at hand was inspired by a host of activities surrounding the Fuller archive. I recall vividly the first telephone call from Jaime Snyder (Fuller's grandson and coexecutor for the Fuller estate) in which we discussed the idea that Fuller's archive might possibly find a home at Stanford University. The telephone call was followed by a series of meetings. Michael A. Keller, the university librarian, and Assunta Pisani, the associate university librarian for collections and services, discussed the prospect of the archive coming to Stanford more seriously with Allegra Fuller Snyder (Fuller's daughter) and Jaime. All parties quickly realized that, indeed, the Stanford University Libraries would be a most appropriate home for the Fuller archive, as Stanford would be able to promote access to the unusually large collection in innovative ways and make the collection all the more available for teaching and research. Shortly after it arrived physically on the Stanford campus, the collection attracted almost immediate faculty interest for teaching purposes. Professor Jeffrey Schnapp offered two consecutive seminars on R. Buckminster Fuller, with students using the archive for their research papers. Shortly thereafter, Michael John Gorman, who was then at Stanford Libraries working with the Fuller archive, began research for a book that was later published as *Buckminster Fuller: Designing for Mobility* (Milano, Italy: Skira Editore, 2005). The Fuller collection has drawn interest from students, scholars, museum curators, and others from any number of fields and disciplines, from design, environmental studies, architecture, and art history to American Studies, and beyond. The interest is both interdisciplinary and international in scope.

Many people deserve recognition for their work on the R. Buckminster Fuller collection at Stanford. We acknowledge with the warmest regard Allegra Fuller Snyder, Jaime Snyder, and John Ferry of the Estate of Buckminster Fuller. We also thank the helpful trio of Thomas Zung, Shoji Sadao, and the late E. J. Applewhite, all of whom were friends and colleagues of Fuller.

The principal editor of the current volume, Hsiao-Yun Chu, has worked tirelessly and steadfastly with all of the authors and the staff from the Stanford University Press on this volume. She worked with the Fuller archive at Stanford for almost two full years and completed projects that have greatly improved both intellectual and physical access to the collection. Hsiao-Yun has been an invaluable colleague, without whom the present work would not have been possible.

Current and former Stanford staff members who worked on the Fuller collection in many and important ways include Glynn Edwards, Mattie Taormina, Hannah Frost, Steven Mandeville-Gamble, Sean Quimby, and Michael John Gorman. A note of thanks also goes to Professor Jeffrey Schnapp, who has been a collaborator with the Libraries on many things Fuller.

Roberto G. Trujillo
Frances & Charles Field Curator of Special Collections
Stanford University Libraries

New Views on R. Buckminster Fuller

Introduction

Hsiao-Yun Chu

In 1999 Stanford University Libraries acquired the enormous archive of R. Buckminster Fuller (1895–1983), one of the most interesting American characters of the twentieth century. Fuller was a self-styled renaissance man whose "profession" proved impossible to summarize according to conventional terms. Fuller himself preferred the phrase "comprehensive anticipatory design scientist," defining for himself a role that was both distributed and nonspecialized. He considered his life to be a cosmic experiment to test whether a single individual could make a difference to the world at large. In his 1982 manuscript entitled "Guinea Pig B," Fuller called himself "a living case history of a thoroughly documented, half-century, search-and-research project designed to discover what, if anything, an unknown, moneyless individual . . . might be able to do effectively on behalf of all humanity that could not be accomplished by great nations, great religions or private enterprise."[1]

In his archive, Fuller left behind tens of thousands of letters and papers, thousands of hours of video and audio recordings, numerous manuscripts, and hundreds of models and blueprints to document his prolific eighty-eight-year experiment aboard "Spaceship Earth."[2] Yet in spite of, or perhaps because of, all this information, there remains a great deal about R. Buckminster Fuller, his work, and his place in history that we have yet to make sense of.

Although several biographies and countless articles have been written about Fuller, both during and after his lifetime, his work has often suffered from lopsided treatment. Some have lauded him as a planetary prophet whose design science work foretold sustainable architecture and nanotechnology; others have dismissed him as a "delirious technician" with a talent for linguistic obfuscation.[3] Between adulation and disdain must lie a balanced picture of Fuller's life and his work. What were the true contributions of his unusual and varied career to science, art, architecture, and society at large? How can we come to a better critical understanding of Fuller's motivations, and how can we use history's distance to assess the significance of his work to the twentieth century and beyond? Our first and best source of information in tackling these questions is the Fuller archive itself; indeed, it was in order to celebrate the archive and its new home at Stanford University that this volume was originally conceived. Thus, we begin with an article that looks at the historical development of the archive, to gain insights into how and why Fuller built this amazing collection over the course of his life.

It is difficult if not impossible to separate R. Buckminster Fuller's work from his personality or, some might say, his persona, and herein lies part of the difficulty in assessing Fuller objectively.[4] His early forays into architecture and design were motivated almost entirely by personal concerns—concerns that would only later mature into a homegrown set of "design science" ethics. His idiosyncratic inventions—ranging from a superefficient car, which crashed horribly on its public unveiling at the 1933 World's Fair, to a low-cost circular house on a mast that could be air-lifted by helicopter, to a proposed dome over Manhattan that would shield the city from snow—have lingered outside historical discourse and been treated almost indulgently as the brainchildren of an eccentric mind. Yet it is important to remember that Fuller's plans, though they bear the hallmarks of his colorful thinking, were responses to contemporary needs for transportation and affordable housing, and Fuller was not the only thinker to propose radical solutions to these problems.[5]

New Views on R. Buckminster Fuller draws on the personal papers, correspondence, and original manuscripts in the Fuller archive to recreate the milieu, both internal and external, that Fuller experienced at different points in time and to look at how his work addressed these circumstances. Indeed, the *work* that we are referring to here is not necessarily limited to architectural artifacts. Actually, Fuller's major contribution to society as a whole may have been ideological, the artifacts serving only to illustrate a continuous discourse that unfolded across his lifetime. Fuller had the unparalleled ability to captivate an audience, to spark the public imagination, and to inspire people to believe that, hitched to the engine of technology, humanity could progress toward a brighter future. Considering the thousands of lectures that he gave around the world, particularly from the late 1960s onward, "Bucky" touched many more lives as a charismatic speaker and public intellectual than as the architect of low-cost housing. He gave some two thousand lectures around the world to packed audience halls, yet only one of his Dymaxion homes was ever built and inhabited. It was through the medium of his performances that low-cost housing and transport became an imminent, if not tangible, reality.

As we mine the Fuller archive to locate the origins of his ideas, what emerges is a complex picture of a man whose long career straddled the entire twentieth century and whose mind-set reflected both the transcendentalist traditions passed down by his revered great aunt Margaret Fuller (1810–50) and the global awareness that characterizes our own times.[6] To shed light on Fuller's mind-set when he began his career in earnest, Barry Katz takes us back to 1927, the year when, according to Fuller's own account, he experienced a great epiphany and resolved to dedicate the rest of his life to the betterment of humanity. Katz recreates a nuanced version of that year, which witnessed not only enormous creativity and self-invention but also the beginnings of a lifelong quest to bridge mechanical ideas with social realities. Howard P. Segal posits that Fuller may have been America's last genuine utopian, a man who was motivated by a vision of how technology could help humankind to realize a more perfect world and who tried to bridge the gap between a dream and reality by laying substantive plans for the Dymaxion houses and cars that would facilitate and complete that utopia.

The Dymaxion inventions of Fuller's utopia were never realized beyond the prototype stage, and countless other design projects, from the mechanical jellyfish to the Fog-Gun shower, never made it to production. However, as Joachim Krausse points out, Fuller's models were important not only as ends in themselves but as a means to further develop his novel ideas and to present them to the public. Krausse examines the complementary relationship between thinking and building in Fuller's work, beginning with the Lightful Houses project of 1928. Through a continuous back-and-forth between thought and development, the Lightful project evolved into the 4D house, which in turn would be renamed the *Dymaxion* house. This unique way of working allowed Fuller to model, crystallize, and further develop his thoughts over the course of his career.

Fuller was fond of the word *precession*, which he defined as "the effect of one moving system upon another."[7] Precession can be thought of as the tangential, even unpredictable, effects of one system's encountering another. Fuller liked to think of himself as a body in motion, influencing, if only ever so slightly, the orbits of those whose paths he crossed and being influenced by them in turn. One way to look at Fuller is to assess the indirect or "precessional" effects of his work. This is in many ways more fruitful and interesting than launching into a critique of Fuller's designed objects, many of which remain frozen in theory as unrealized patents.

As a tireless teacher and lecturer, Fuller encouraged lateral thinking and inspired others to pioneer new developments. Because he moved across a myriad of fields, traces of Fuller are found in the most unlikely places. Fuller's work on geodesic structures is found in the seams of soccer balls and the struts of children's jungle gyms, and his dome for Expo 67 still overlooks the Montreal skyline. His design science work sometimes presaged other discoveries, notably the structure of the carbon-60 molecule, or "buckyball," in chemistry by several decades.

Just as Fuller affected others, he was also affected by them. While Fuller's structures have often been assumed to be the futuristic creations of a visionary mind, his ideas were in fact neither ahistorical nor wholly without precedent. His own work and thinking drew both on past traditions and the work of his contemporaries. Several authors place Fuller's work in context by showing how it related to the larger spheres of history. Claude Lichtenstein investigates how Fuller's Dymaxion house answers to a particularly American vision of the home as described decades earlier by the Beecher sisters in *The American Woman's Home.* David Nye's article discusses the concept of energy and how it is manifest in Fuller's designs across the years, from the service core at the center of his Dymaxion house to his proposals for worldwide energy grids that would share electricity around the world. Jonathan Massey discusses Fuller's unique aesthetic, born of his concerns with geometry, time, space, and economy, relating it especially to Claude Bragdon's earlier ideas of projective ornament. Maria Gough investigates the question of attribution, suggesting that although Fuller's architectural ideas as exhibited at New York's Museum of Modern Art in 1959 added new grist to the modern architectural discourse, he may have left some unpaid debts among his former students, many of whom, according to Gough, "were not always properly credited for their often fundamental role in his design innovations."

Fred Turner and Felicity D. Scott revisit the 1960s and 1970s, when Fuller experienced a surge of popularity with the counterculture. Turner recounts how, in an ironic twist of history, the technocratic septuagenarian who had once built shelters for the U.S. military became an unlikely ally in the young people's struggle for a less-bureaucratic world. Fuller helped to restore their faith in technology by indicating how small-scale technologies could help them realize their dreams of social change. Likewise, Scott shows us how Fuller's World Game and his domes provided hope and inspiration for so many disenchanted youth, though it ultimately failed to solve the political and social realities that were the real source of their discontent.

Fuller's life and work continue to generate more questions than answers. Thus, it is fitting that we end by pondering Reinhold Martin's suggestion in "Fuller's Futures": was Fuller a postmodernist? Within his lifetime Fuller suggested an entirely new relationship between human beings and the universe, where local concerns give way to universal frames of reference; yet, as happens when wandering in an infinitely self-similar fractal, we risk losing our bearings completely in this new reality.

With its seemingly endless boxes and files, the R. Buckminster Fuller archive puts us in a dilemma so characteristic of the postmodern information age: we find ourselves awash in superhuman amounts of information but must use human means to navigate and make sense of it. As we dive into a vast archive that has been all but forgotten for decades, this volume marks the beginning of that journey.

1 Paper Mausoleum

The Archive of R. Buckminster Fuller

Hsiao-Yun Chu

One first opens a box from the R. Buckminster Fuller archive with the voyeuristic anticipation of peeking into the personal papers of a misunderstood genius. But this expectation begins to fade as one becomes lost in a sea of ephemera: receipts, thank-you notes, business cards, form letters, and newspaper clippings. For a researcher the promise of absolute completeness quickly becomes tempered by the frustrations of excess. The Fuller archive comprises some fourteen hundred linear feet of material and seventeen hundred hours of recordings. It is by far the largest archive of a single individual ever processed at Stanford, if not in the country, and it would take years, if not decades, to go through this collection. Although it has been consulted by researchers and marveled at by visitors, the archive has not been treated in any depth as a research subject per se. Yet it is a central phenomenon in Fuller's story, arguably the most important "construction" of his career, and certainly the creative masterpiece of his life.

Fuller had multiple reasons for organizing and maintaining his massive collection—some obvious, stated, and professional; others intrinsic, unstated, and personal. As a young man he foraged into his own family history to complete a detailed record of his ancestry; in parallel, he began to save almost every piece of paper that crossed his desk in order to construct an exhaustive chronology of his own life. The archive became a repository of technical information, a personal library that lent intellectual credibility to his unconventional design projects and helped to buffer the sting of his early intellectual failure, having been firmly expelled from Harvard as an undergraduate. Increasingly, the archive became more polyphagous, ingesting not only every piece of paper touched by Fuller, in chronological order, but newspaper clippings, recordings of speaking engagements, and student projects related to his work. By the time of his death in 1983, the archive contained tons of papers, thousands of hours of audio and video footage, and hundreds of models and assorted artifacts. By investigating the complex and changing set of reasons that compelled Fuller to maintain this huge collection across the decades of his life, often at significant cost and burden to himself, we will come to a better understanding of Fuller's character and gain a greater appreciation for how the archive both authorized and personified its maker.

Organization of the Fuller Archive

At the heart of the Fuller archive is the "Dymaxion Chronofile," a chronological arrangement of outgoing and incoming personal and business correspondence, receipts, greeting cards, business cards, and the like. Fuller began collecting documents at an early age, including newspaper clippings, letters, and other items that evidenced "the wonders of 'modern' technology."[1] He began to formally order his papers in 1917 and, some years later, would christen his collection the Dymaxion Chronofile.[2] According to Fuller, his choice of a chronological arrangement was influenced by his service (beginning 1917) in the U.S. Navy, when he realized the value of keeping accurate operating records organized in a fashion that would be easily accessible.[3] A chronological arrangement obviated the need to remember precise names or details. To find a record, one need only recall the approximate time in the past that an event had happened; by working forward and backward from that point, the information could be located fairly quickly. Time provided an unbiased, linear metric for arranging records of events.

Fuller presented the Chronofile as an objective and accurate file documenting the life of a human being starting from the Gay Nineties and moving forward into a very different and much faster-paced century. "I decided to make myself a good case history of such a human being and it meant that I could not be judge of what was valid to put in or not. I must put everything in, so I started a very rigorous record," said Fuller in 1962. "I thought it might be interesting if I took . . . everything, not just culling out the attractive aspects of my life, but really keeping the whole records—most of which was not so attractive, and putting it all into chronological order."[4] Although the Chronofile was the earliest organized portion of Fuller's papers, over the decades his archive grew to include manuscripts, newspaper clippings, video and audio recordings, and photographs. With the exception of photographs, most of these sections were basically chronologically ordered as well.

The democratic way in which Fuller treated the ephemera of his life—each letter, scrap, pamphlet, and clipping—is admirable in its objectivity. Still, one wonders what compelled this man to save everything, especially considering that Fuller himself conceded that much of the material was not "attractive." Furthermore, the archive seems to fly in the face of Fuller's insistence that designers and humankind in general must move toward "ephemeralization," namely "the principle of doing ever more with ever less, per given resource units of pounds, time and energy," in order to reach the necessary efficiencies to support all humanity.[5] How could a person who advocated stringent resource management and judged a building based on its volume-to-weight ratio justify his lifelong maintenance of a gargantuan file that was neither light nor efficient? At the time of Fuller's death his former archivist estimated the weight of the archive to be ninety thousand pounds.[6] Ironically, in all its massive glory, the archive itself suggests quite another definition for "ephemeralization" than that intended by Fuller—namely, "the willful construction of a seemingly vast collection of ephemera."

Origins of the Archive

Professionally and publicly, Fuller presented his archive as an invaluable research collection. "The R. Buckminster Fuller collection of life's work constitutes a vast amount of *Raw Material*, in many different physical forms," wrote Fuller and his staff in a document describing the archives in 1965.[7] "I decided in 1917 to contribute to the scientific documentation of the emergent realization of the era of accelerating-acceleration of progressive ephemeralization."[8] According to his accounts, Fuller had undertaken this large archiving task primarily for the benefit of humanity, to create a comprehensive repository of information documenting the revolutionary changes wrought by the twentieth century. However, given that the archive is clearly specific to the life of R. Buckminster Fuller himself (as opposed to that of any other human being born in 1895), can it really be said to be objective? Embedded in the archive is the assumption that Fuller's life per se is worth representing. The 1965 document is, in fact, woven on the warp of Fuller's ego. "The collection could be likened to the papers of a renaissance man[,] a true comprehensivist such as Leonardo Di Vinci [*sic*], but for one large factor; no one knows what of Di Vinci's papers were lost or destroyed because they were thought insignificant in an age of many comprehensivists. Everything remains of Fuller's work because he saw to it."[9]

That the supposedly objective archive was clearly subjective belies one of Fuller's peculiar attributes; the coexistence of hubris and humility. Fuller was a self-made man and a tireless self-promoter, but he tended to disarm people with his unselfconscious, even self-deprecating, style. He often referred to himself as "Guinea Pig B," a test creature whose life was an experiment aimed at discovering what one individual could do to help all of humanity; yet the idea of saving every last scrap that crosses its path in order to document this experiment is a rather grandiose conception for a guinea pig. This is not to discount the value of this collection; Fuller's archives do indeed record significant events of the twentieth century as seen through the lens of his activities, and he corresponded with many interesting people.[10] Still, the idea that the archive was an "objective" lab notebook must be taken with a grain of salt. It is perhaps more objective to say that the archive was a (self-)conscious arrangement, formed with the understanding that every piece of paper that came into Fuller's world must be filed and saved. The amassing of the archive was a lifelong creative act that can easily be seen as a masterpiece of conceptual art. In the end the archive is both "everyman's" collection (in that it is made up of the common, mundane ephemera of twentieth-century life) and the work of a specific individual, who hallmarked the collection by the very act of creating it.

It is telling that the first section of the archive, in keeping with the organization imposed by Fuller and his archivists, deals with Fuller's family history. Chronologically, it makes sense that these materials would come first; family history was an important psychological foundation for Fuller. He drew confidence from the stories of his ancestors, several of whom were notable individualists. His great, great, great, great, great grandfather, Lieutenant Thomas Fuller of the British Navy, came to America on leave in 1638 and settled in New England, excited by the possibilities of the new country.[11] Fuller's Massachusetts ancestors included four generations of Harvard men (an unbroken line until young "Bucky" came along), among them influential lawyers and chaplains in the Boston area. Fuller was especially interested in the story of his great aunt Margaret Fuller (1810–50), a prominent New England transcendentalist who, together with poet Ralph Waldo Emerson, founded the literary magazine *The Dial.* She moved to Europe in 1846 and became the first female foreign correspondent to work for a major newspaper, the *New York Tribune.* The family history files in the Fuller archive include letters that Fuller exchanged with archivists and local historians; photographs of portraits of his ancestors obtained from the archives of churches in Cambridge, Massachusetts; and the like. Family history remained one of Fuller's passions throughout his life. In 1927, on the birth of his daughter, Allegra, Fuller compiled and privately published a detailed family chronology called the *Record of the Direct Parentage of Allegra Fuller*, hoping to instill in his daughter the same pride of lineage that he had. In 1983, the year of his death, Fuller hired the Debrett Ancestry Research to research the life of Lieutenant Thomas Fuller in more detail.

On this foundation of family history Fuller began to build his paper bulwark. He began systematically saving and filing his papers and correspondence in 1917. The name *Dymaxion Chronofile* was likely given to this set of papers in or after 1929, when the word *Dymaxion*, a combination of the words *dy*namic, *max*imum, and tens*ion*, was coined for an exhibition of Fuller's model home of the future at Marshall Field's department store in Chicago.[12] Initially the Chronofile consisted of various personal and business papers bound into handsome, leather-backed volumes.[13] Eventually, it became too expensive and space-intensive to produce volumes; from about 1940 on, many of the papers were left unbound, in folders, and newspaper clippings were kept separately from correspondence. From about 1973 to 1983, carbon copies of Fuller's outgoing correspondence were kept in a separate file. Audio- and videotapes and film reels were also added to the collection. In its current state the Fuller archive consists of twenty-three distinct series, the largest and most important of which remains the Chronofile.

The Archive as a Support

Fuller recounted many times over the defining moment of his life. It was the winter of 1927, and he had suffered serious professional and financial failures and the death of his first daughter, Alexandra. A new daughter, Allegra, was born in August 1927. Unemployed and desperate, Fuller felt that he might be worth more to his young wife and daughter dead than alive; he had a good life insurance policy, after all. Fuller wandered to the icy shores of Lake Michigan, ready to take the last swim of his life. We may never know fully the actual details of this lonesome struggle because as the story was told and retold over the years, it attained mythic proportions. From the depths of his despair, Fuller describes being lifted several feet off the ground by an invisible force. He heard a voice telling him he had no right to eliminate himself; rather, it urged him to think the truth and follow it. In the wake of this spiritual revelation, Fuller nominated himself as the test case for an experiment aimed at discovering what one relatively average individual could do to better the world for the future. After his epiphany Fuller lunged into an almost desperate frenzy of activity to fulfill his new mission. The late 1920s were an extraordinarily prolific period of writing, sketching, and personal work, wherein one finds the seeds of much of Fuller's later work (fig. 1.1).

Although Fuller rebounded from the Lake Michigan experience in a highly charged creative mode, it is important to note that the revelation that changed the course of his life emerged from an abyss of self-doubt and fear that had been with him, in varying intensities, for much of his life. Fuller was plagued by deep insecurities, in part the product of his not-so-carefree youth. Until the age of four Fuller suffered from undiagnosed but severe nearsightedness and experienced the world as a fuzzy fog of moving forms. His images of reality never matched up with those described by his siblings, so he invented colorful

Figure 1.1.
R. Buckminster Fuller in New York, c. 1929.
Source: Special Collections, Stanford University Libraries.

tall tales about what was going on around him, believing that his siblings were doing the same.[14] When he finally received prescription eyeglasses at age four, he experienced a completely new reality that rendered his past a false impression. In 1907, when Fuller was twelve, his father suffered a debilitating stroke, and the young adolescent watched him turn from a healthy, successful businessman to a frail, disconnected invalid. Undoubtedly, the loss of his father three years later was a traumatic experience for the young Fuller. He entered Harvard in 1913, but his first year was quite frustrating as he failed to gain entry into any of the prestigious social clubs. To compensate, Fuller tried to distinguish himself athletically and joined the football team, but a knee injury he suffered during practice took him out of the game. He was expelled from Harvard in 1914 owing to poor performance but readmitted several months later, only to be expelled for the second and final time in 1915. In 1922 his first daughter, Alexandra, died of spinal meningitis, and Fuller, who was scraping together a living, felt that he was partly to blame for failing to provide a better home environment for her. He was prone to drinking heavily over the next few years and never quite overcame his regret. In 1926 Fuller lost his job as head of the Stockade Corporation, a building firm that he had helped his father-in-law, James Monroe Hewlett, to found four years earlier.[15]

In general, Fuller was a gregarious character and seemed to bounce back from adversity time and again, offsetting failures with successes. The 1927 incident at Lake Michigan was the closest he came to succumbing to his feelings of self-doubt. Although an invisible spiritual force had pulled him back from the brink of self-destruction that night, he also had ample evidence to prove that his life had not been a complete failure. Already in 1927, his growing collection of papers had begun to assuage Fuller's pervasive fears: "it [his collection of papers] persuaded me ten years after its inception to start my life as nearly 'anew' as it is humanly possible to do," wrote Fuller nearly forty years later.[16] The constant task of recording events about himself had served to objectify them, not as fears or successes but as part of the experimental process that he, as Guinea Pig B, had undertaken. Although in conventional social terms he may have been a failure, this became irrelevant within the larger scheme of the lifelong scientific experiment, as the conducting of the experiment itself was a success. His burgeoning laboratory notebook suggested to him that he usually achieved more when he followed his own lead rather than when he tried to conform to "everybody else's pre-fabricated credos, educational theories, romances, and mores."[17] Thus it was the Chronofile itself that provided evidence for him to reshape his life and reaffirm his priorities in his darkest hour.

Furthermore, although the Chronofile was "scientific" in its conception, it held for Fuller enormous personal and emotional value. After all, it contained not only business correspondence and technical notes but also love letters from his wife, photographs, and notes from dear friends. Symbolically, through papers, words, and images, it stood as a powerful monument to his own self-worth.[18] As such, it was a real source of security and self-esteem.

Throughout Fuller's career the Chronofile supported him not only emotionally but also professionally. In effect, Fuller was building a sizable "research library" around himself, which put him on par with the academic world that had once rejected him. The Chronofile lent credibility to his work and helped to validate its importance. When critics questioned his seemingly far-flung ideas about housing and transportation, he would point to his mounds of evidence, data collected, letters exchanged, and work done. Indeed, it was difficult to argue that Fuller's "scientific" methods might be flawed, when he had forty-five tons of papers in tow to the contrary.

After 1927, knowing that he was not a success by conventional standards, Fuller chose his own metrics by which to measure his worth; he drew his own stock charts, not according to amounts of money but amounts of paper. From

1947 to 1981 Fuller's office staff kept a "Dymaxion Index," a bibliography of published news items about Fuller and his work that was updated by his office on a regular basis. The index also contained schedules showing Fuller's lecture activities year by year and chronological lists of the projects he had worked on. The "Dymaxion Index" was used to create graphs such as "Major News Items About R. Buckminster Fuller," which shows progressively larger peaks over time into the hundreds (fig. 1.2). Such graphs were updated regularly, based on the bibliographic entries in the "Dymaxion Index." Similarly, Fuller kept charts of the "Dymaxion Chronofile Correspondence." One example dating from 1975 has a solid line showing about ninety-four hundred incoming and outgoing letters in the year 1974, and a dotted line showing a cumulative lifetime total (1895–1975) of close to one hundred thousand items (fig. 1.3). Fuller's office staff would often include copies of such charts in letters to newspapers, publishers, and universities as visual proof of his social and intellectual importance.

The skyrocketing "correspondence" totals in Fuller's charts make an impressive point. These numbers should be revisited, though, in light of the fact that Fuller had his office staff file and reply to even the most mundane incoming letters and to keep a carbon copy of the reply. The archive contains hundreds of inquiries about where one could purchase Dymaxion maps, for example, and carbon copies of the standard responses. Alumni letters soliciting contributions to the Harvard Fund were saved, as were the thank-you notes that he received from the Alumni Association. During the 1960s Fuller's office staff contacted literally thousands of manufacturers and suppliers to compile an inventory of the world's resources. Each outgoing request for information, and each incoming cover letter and brochure or catalog, would be saved in the archive. In other words, the correspondence files of the archive are more impressive in number than in actual content.

While the early decades of the Chronofile (1920s–30s) are dominated by personal letters from his wife, Anne, and business items, the later years of correspondence (1960s–80s) contain thousands of letters that Fuller himself neither personally read nor wrote. The archive ballooned over time, as it included not only Fuller's personal work but also projects that his office staff were working on. As such, the archive became large enough to shield not only Fuller but also his staff from the skepticism that invariably accompanied their projects. In 1970, for example, Medard Gabel, a student of Fuller's who worked tirelessly on the World Game project, spoke passionately to his audience, "'We can make everybody in the world a success by 1980 and we have the proof to back it up,' he said. 'So when we say that air pollution could be eliminated from Spaceship Earth by 1980, that's exactly what we mean.'"[19] The "proof" that Gabel was

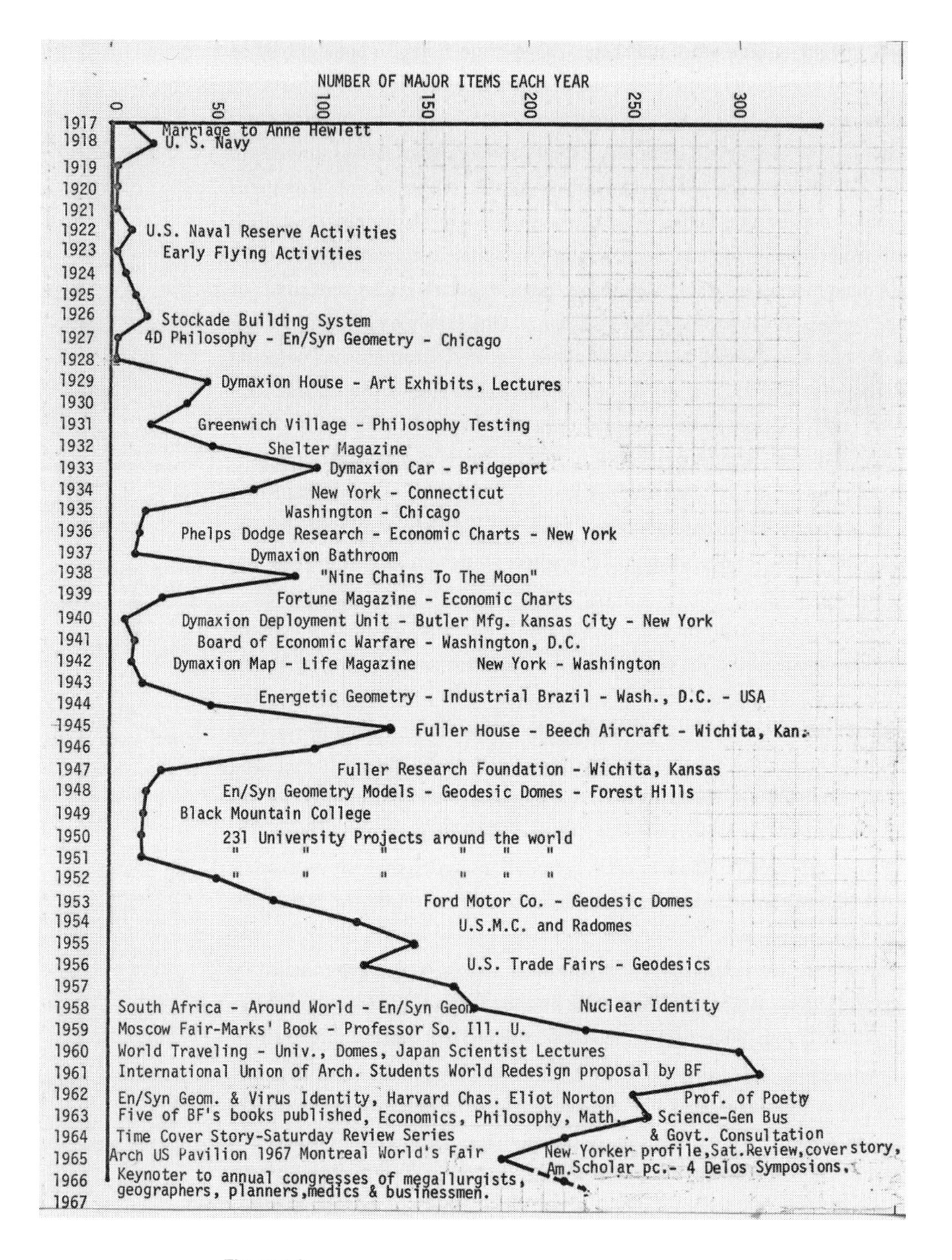

Figure 1.2

Major News Items about R. Buckminster Fuller, compiled by Fuller's office staff, 1917–67.

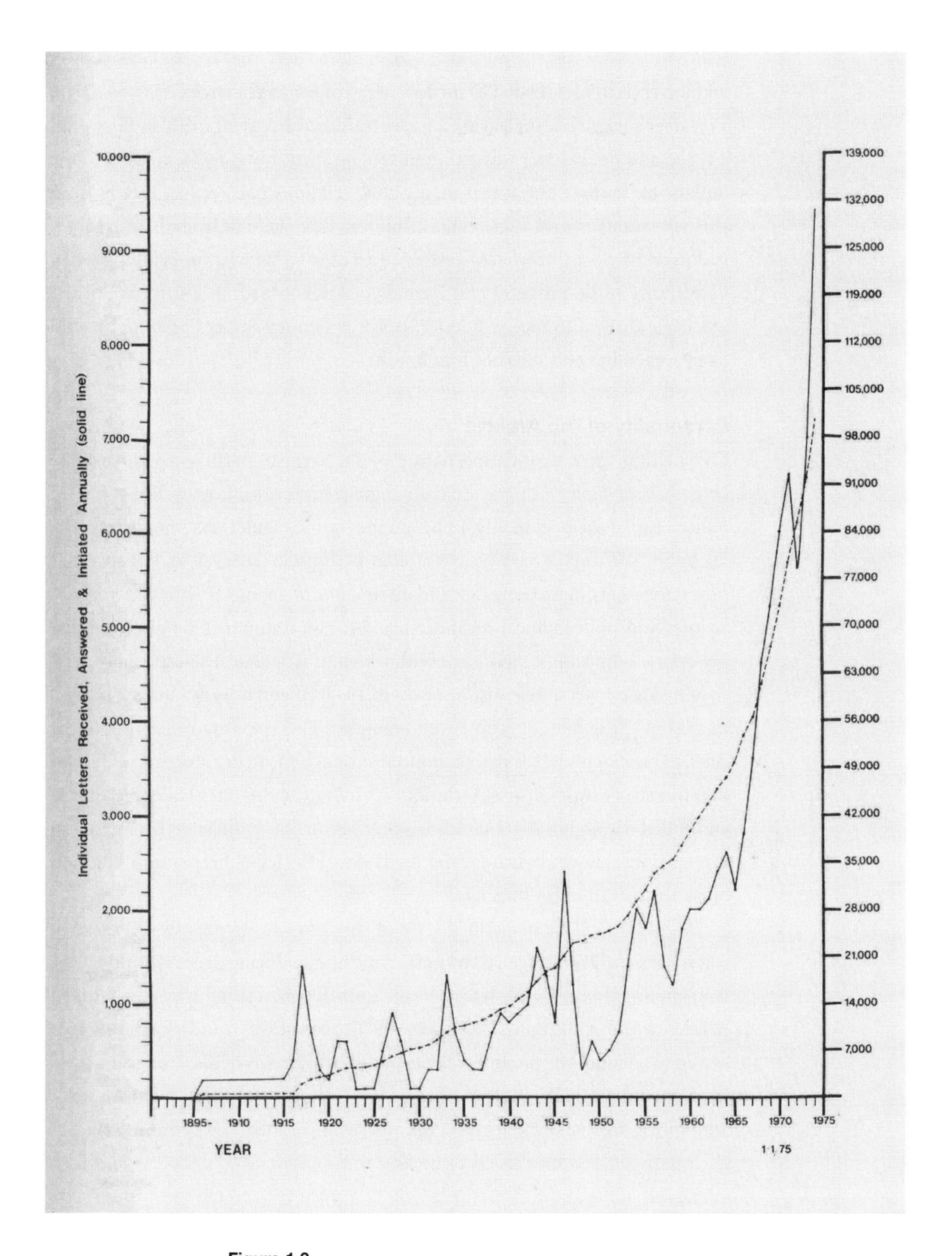

Figure 1.3

Chronofile correspondence, 1895–1975.

 Source: Special Collections, Stanford University Libraries.

referring to here was the piles of papers generated by the World Game project and the endless files related to the inventory of world resources.

Fuller's insistence on leaving a paper trail and answering every inquiry with a typed and dated letter was an important strategy, not only for adding to the bulk of his archive but also from a public relations perspective. Letters from curious strangers and eager schoolchildren alike were answered courteously and promptly—a throwback, perhaps, to Fuller's Victorian upbringing when letters were to be properly and formally acknowledged. In return, recipients would be thrilled to receive a letter from the famous Bucky Fuller and would likely remember and speak of him fondly.

Personality of the Archive

Considering the ever-inflating nature of the archive, particularly during the latter half of Fuller's career, we cannot help but notice how well it reflected Fuller's expansive personality. In his public lectures Fuller was known to speak for hours at a time, tirelessly, in a stream-of-consciousness style. His speeches mixed together engineering talk and discussions of geometry with personal anecdotes and philosophical meanderings. "Always unhurried, he gives you the impression of having a lot of time with which to articulate his perpetually multiplying ideas," wrote *Science Illustrated* in 1944. "If you miss out on one phrase, the rest of the sentence becomes unintelligible. This leads many listeners to the conclusion that his talk is mainly nonsense, or at best, an engineering version of the stream-of-consciousness technique."[20] Indeed, sometimes Fuller became so embroiled within his own marathon speeches that he subjugated his biological urges; he was known to urinate discreetly down the leg of his pants rather than break the thread of his musings.

Although Fuller's unchained style of speaking was frequently criticized, it was surprisingly well-received by university hippies during the 1960s and 1970s, who apparently enjoyed listening to the quixotic guru spin his personal blend of science and philosophy. Ostensibly, the use of various mind-expanding aids served to enhance the totality of the experience for many of his young admirers. In January 1975 Fuller gave a series of lectures over the course of two weeks, amounting to a historic forty-two-hour session called "Everything I Know." Each individual lecture lasted some four to five hours, and people would drift in and out of the auditorium while Fuller rambled on. Around the twenty-hour point, Fuller spoke of the "total experience" that his audience would have if they stayed for the entirety of the "Everything I Know" marathon. "I think we're going to have the total experience. If we get to the end of the time, I'm quite certain that I'm not going to be withholding from you some of the things that I feel are all this important interrelatedness, because I do come into you time and again with new kinds of thrusts, and yet you find everything getting back into the same fundamental world. It really gets more and more thrilling."[21] With similar sexual undertones Fuller describes how his expansive ideas will fertilize his audience, until eventually everything and everyone is interrelated.

These themes of continuous expansion and interconnectedness are hallmarks of Fuller's "comprehensive" thinking. The idea that everything in the natural world is interconnected, and that by understanding local activity one can uncover the fundamental principles of the universe, is central to Fuller's mission as a "comprehensive anticipatory design scientist." Fuller described himself as a nonspecialist who addressed the universe as a whole and who had the rare ability to think objectively and on a large scale. Whereas other people concerned themselves exclusively with local projects and problems, Fuller thought in "comprehensive" or universal terms. His expansive speeches, which spun out to discuss integrity, humanity, and ultimately the function of "man in [the] universe," seemed to transcend the mundane limits of "realistic" thinking. With eyes closed and hands clasped, as he often faced his audience, the stubby, impassioned Fuller resembled an oracle whose cryptic, rambling speeches suggested some connection to a larger spiritual force (fig. 1.4). Presumably, this expansive attitude also stoked Fuller's sense of self-worth, because it claimed for him the "prerogative(s) of limitless self-extension, what we might call 'cosmic significance.'"[22]

Likewise, the Fuller archive shared the comprehensive, expansive nature of its creator, providing tangible proof of his "cosmic significance." In 1965 Fuller wrote that "the subject matter of the collection covers a very broad spectrum, practically all areas of human knowledge."[23] During the 1960s and 1970s Fuller devoted most of his time to giving lectures around the world.[24] He maintained an incredibly ambitious travel and lecture schedule, pollinating the globe with his ideas and eventually becoming a household name. At home the archive likewise continued to proliferate, ingesting papers and swelling in size under the stewardship of his cadre of administrative assistants.

Fuller's insistence on comprehensiveness and his time-dependent filing system amount to a totalitarian system of control over the arrangement of the archive that denies any assignment of value, in spite of the fact that a majority of the papers are of only nominal interest. At the same time, the archive seems to fulfill Fuller's pet notion that by investigating local phenomena, one can discover universal truths. The archive takes one life and documents it as thoroughly as possible; this microcosm validates the general observation that "life is made up of a great number of small incidents and a small number of great ones."[25]

Figure 1.4 R. Buckminster Fuller lecturing. Photo by A. Mikkelsen. Source: Special Collections, Stanford University Libraries.

Fuller's Legacy

We have seen how the Fuller archive provided emotional support and academic credibility for its creator, how it embodied his expansive nature, and how his insistence on comprehensiveness led to an archive of immense proportions. Besides the professional and the personal, there were also practical considerations surrounding the collection. Throughout the years its care and feeding created considerable expenses for Fuller himself. In 1959 Fuller was invited to become a research professor at Southern Illinois University at Carbondale (SIUC) by then-president Delyte Morris. Fuller brought a certain amount of prestige to the university; divested of formal teaching responsibilities, he had carte blanche to pursue research and to travel as much as he wished. But one of the main perks of his position at SIUC, where he worked from 1959 until 1970, was to establish a home base for his burgeoning archive. The university agreed to pay for storage of Fuller's papers and for a half-time salaried assistant to administer the collection.

Eventually, SIUC was supposed to become the permanent home of the Fuller collection. As early as 1960, he wrote to Dr. Ralph McCoy, SIUC's director of libraries, that he would be willing to give to the library (1) the Dymaxion Chronofile; (2) published and unpublished manuscripts; (3) drawings and photograph files; and (4) the complete collection of all published clippings concerning his work, provided that certain stipulations were met. Among others, the conditions were that the papers be housed under lock and key in the Rare Books division of the library; that the Chronofile would not be accessible to the public (owing to the "intimate, personal" nature of some of the documents within) until some specified future date; and that microfilms be made of all the clippings in the collection, and a copy of the masters given to Fuller.[26] The collection as described was valued at $15,000; Fuller intended to use this figure for a tax deduction after making the final gift.

In a letter dated May 5, 1965, Dr. McCoy confirmed that Fuller's papers were being "housed in the vault of the Rare Book Room, which is kept locked at all times and to which only the rare book librarian has the key."[27] Clippings were being microfilmed, and approximately eighteen hundred drawings and prints had been cataloged as of May 1965.[28] Even so, Fuller's papers extended beyond the walls of the library. Fiscal records from 1964 to 1965 show that the university was expending $3,000 per year on administrative salaries and $2,700 on space rental to accommodate Fuller's additional papers and project files.[29]

The collection remained in SIUC's Morris Library until 1971, but, in spite of the original agreements, it was never actually gifted to the university. In 1970 Fuller's relations with SIUC, which had been his professional home longer than any other institution, began to sour. The Vietnam War, and especially the Kent State riots, in which four antiwar students were killed by national guardsmen, had made the students increasingly restless. Students at SIUC responded with riots of their own, forcing the chancellor to shut down the university for several days starting May 12, 1970. President Delyte Morris, who had brought Fuller to Carbondale, faced widespread criticism for having handled this emergency poorly, and he stepped down a few months later. Many of Fuller's longtime associates, including chief counsel John Rendelman and Dean Bernard Shyrock, also retired or left the university at this time. By 1971 the new administration, which had suffered great financial setbacks because of low enrollment, did not have the discretionary funds to support Fuller's plum research professorship.[30]

Adding insult to injury, in 1971 Fuller had a major falling-out with Tom Turner, an administrator who had been chosen by the university to oversee and administer Fuller's projects a few years earlier. In various memos Turner commented on the "bizarre antics" of Fuller's assistants and their "total disregard of or lack of understanding of academic protocol."[31] He also wrote of Fuller's problems "accepting guidance from administrators" and his "need to 'unsettle organizational authority and structure.'"[32] Fuller went straight to the university's board of trustees, asking that Turner be reassigned immediately, and was offered an apology by the board. Even so, this incident confirmed Fuller's conviction that SIUC was now being directed by "a rump committee of tenured professors . . . complete strangers to me."[33]

In the summer of 1971 Fuller decided to move to the SIU campus at Edwardsville, some one hundred miles north of Carbondale, where he believed he would be more welcome. When Fuller left, he took his papers with him, reneging on his earlier promises to give them to SIUC's Morris Library. There was, at that point, still some hope that the papers would be given to the library at SIU, Edwardsville. But in 1972 Fuller wrote a long letter to Dr. McCoy, formally withdrawing his papers from SIU. Fuller claimed that the university had failed to microfilm his clippings collection as promised and had stored materials inadequately; he preferred to store his materials in his own office in Edwardsville, where he could exert absolute lifetime control over his papers.[34] Thus the papers were released and transferred to storage spaces near Fuller's new office in Edwardsville.

Although political flaps were part of the reason why Fuller withdrew the gift of his papers from Southern Illinois University, there is evidence that his thinking had changed regarding the desirability of using the archive for a tax write-off. After all, the amount that he had invested over the course of his life to store, administer, and manage the archives was far more, dollar-wise, than he would recoup from a one-time tax break. In a 1971 letter Fuller noted that "I have on my own personal payroll one full-time and three part-time people, under Mr. Klaus' supervision, who devote full time to the maintenance of my papers."[35] He often paid for storage space out of his own pocket. Fuller's income during the 1970s consisted of lecture fees, his university income, and some money from licenses and publications. These funds were quickly disbursed to pay for his staff, project supplies, rent, and travel. Although Fuller and his family lived quite well, he was by no means as wealthy as he was famous. Fuller had always posited his papers as a research collection of educational and historic value to future generations of humanity at large, but in his later years he began to think seriously about its value to a narrower subset of posterity: his own.

In a letter dated April 5, 1971, Fuller's assistant, Ed Applewhite, noted that Fuller was mindful of the many financial sacrifices that his family had made throughout the years; for example, he "had to borrow from his wife and has not been able to amass any capital." The archive seemed to promise some recourse. Applewhite wrote that Fuller believed the manuscripts in the archive were now quite valuable. "He said one manuscript might bring in enough money to put his grandson, Jaime, through college and that he absolutely had to provide against this kind of need."[36] In 1972, after he had withdrawn the gift of his archives to SIU, Fuller wrote, "I have invested my entire life earnings and all inherited finds in research and development initiatives. My only economically supportive legacy to my wife, daughter, and grandchildren is that of the royalties accruing to publication potentials existent in my archives."[37] By reclaiming the archive, he also reclaimed his legacy.

In 1979 Fuller considered selling his archive outright, this time valuing it at far more than the $15,000 he had agreed on with SIUC some years earlier. Apparently, there had been some interest on the part of the University of Rochester Library to acquire the papers. Fuller enlisted Charles Sachs, a well-established dealer of rare documents who worked at the Scriptorium in Beverly Hills, to represent his archive, which would be sold according to Fuller's valuation. Not surprisingly, perhaps, Fuller measured the value of his archive in terms of the weight of gold. He wrote to Sachs:

> I first gave him a figure of $1 million in terms of the value of gold. By the time the University of Rochester Library wrote me, I had to give them a figure of $1.5 million. For the same reason, I had to give you a figure of $2 million. The fact of the matter is that the value of my archives is always increasing, not decreasing as is the American dollar. The day I was talking to you, gold was selling for $375 per troy ounce. At that rate, the price of the archives would be 5,300 troy ounces of gold, so that will remain the price I have given you.[38]

Sachs apparently had trouble working with this kind of valuation and as of six months later had not gotten a buyer. But Fuller defended himself:

> The more the gold price, the more valuable the archives for they prove that the metaphysical "know-how" is more valuable than "what." In due course (maybe half a century) the value of *my* archives—not archives in general—will be far greater than that of tons of gold for I do have the know how for humanity to make it here on board Spaceship Earth, here in the universe. Gold can't buy that![39]

Apparently, Sachs was unable to get bids that satisfied Fuller's ambitions, and the papers were not sold during Fuller's lifetime. After Fuller's death in 1983, the bulk of his papers were boxed up and kept in a rented storage space in Santa Barbara, California. A part-time archivist responded to inquiries about the papers and would help the occasional researcher to navigate through the boxes. However, the vast majority of materials remained in cardboard boxes, occluded by other boxes, untouched for about a decade. The storage space was not air-conditioned, and there were minimal environmental controls. The cost of renting the space was a constant drain, and the staffing was not adequate to support a research archive of its size. In 1999 the collection was finally sold to Stanford University Libraries; the bulk of the money from the sale went to Fuller's family, and the copyright remained under the control of the Estate of Buckminster Fuller, a small organization representing the Fuller family. Thus, Fuller's hope of providing for his descendants came to pass, bringing the family a reasonable sum of money and copyright control ad infinitum. Stanford assumed the responsibility of sorting through, arranging, preserving, and cataloging every last item in the vast archive, a process that took a little over six years to complete.

According to the Buckminster Fuller Institute, some three hundred thousand geodesic domes have been built around the world, including the famous dome built for Expo 67 in Montreal and the 530–foot Tacoma Dome in Tacoma, Washington. Fuller built the Dymaxion house, the Dymaxion car, and numerous dwelling units for the U.S. military. But notwithstanding his engineering work, the most important, glorious, and valuable construction of R. Buckminster Fuller's entire life was his archive. With the Dymaxion Chronofile at its heart, the archive documents every activity and every encounter of every day of Fuller's adult life; he created the archive in his image and in his name, and after his death, it became the most accurate reflection of his history. It was his advocate and his ally; his comprehensive compendium; his life and his legacy.

1927, Bucky's Annus Mirabilis

Barry M. Katz

2

I

In the last months of 1927, walking morosely along the shores of Lake Michigan, brooding over the loss of his business and the birth of his second child, R. Buckminster Fuller underwent an experience that has all the trappings of a religious conversion. Here is a passage—drawn almost at random from his voluminous notes, chaotic manuscripts, didactic letters, interminable speeches, contentious interviews, and rambling publications—in which he captures the moment that would define the rest of his extraordinary career: "Standing by the lake on a jump-or-think basis, the very first spontaneous question coming to mind was, 'If you put aside everything you've ever been asked to believe and have recourse only to your own experiences do you have any conviction arising from those experiences which either discards or must assume an *a priori* greater intellect than the intellect of man?'"[1]

The answer to this epistemological dilemma, which recalls the *Confessions* of Augustine, the *Meditations* of Descartes, and the *Autobiography* of John Stuart Mill, came to Bucky in a blazing revelation. The human intellect, when coupled with the range of possible human experience, is sacred: "Apparently addressing myself, I said, 'You do not have the right to eliminate yourself, you do not belong to you. You belong to the universe.'"[2]

Fuller would return again and again to his crisis of 1927 to create a picture of nothing short of death and resurrection. At one moment he is gazing out onto the frigid waters of Lake Michigan, mired in self-doubt and seriously considering putting an end to it all. In the next he is reborn, confident of his place in the universe, rising bravely to his self-appointed, world-historical destiny. Selflessly he will abandon the pleasures of work, career, and income in pursuit of what he calculated would be a thirty-year experiment in self-actualization, "designed to become effective in 1952." He will shun the worldly preoccupations of architects, engineers, and educators and commit himself wholly to the articulation of a new design science. He will renounce language itself, the proximate source of all error and confusion: "It was very tough on my wife, but I decided I was going to try to hold a moratorium on speech. . . . So for approximately two years I didn't allow myself to use words."[3] But words would return to him—with a vengeance—and they guide us toward an insight into 1927, Bucky's year of wonders, his annus mirabilis.

II

The path that brought Fuller to his moment of crisis and redemption has been retraced by many visitors to the Dymaxion world, and he himself cultivated his personal narrative to what can only be described as mythic proportions. Indeed, it in no way detracts from the gifted inventor of the Dymaxion car and the geodesic dome to say that Bucky's most inspired and inspiring invention was himself. The continual recitation of the events of 1927—"In 1927 I committed all my productivity potentials toward dealing only with our whole planet Earth"; "I began to look at environments in 1927"; "In 1927 I undertook the search for a reliable means of estimating world population figures"; "I set out deliberately in 1927 to be a *comprehensivist*"—served as a template that imparted to his career the regularity of an icosahedron and the inevitability of an octet truss.[4]

Over the course of innumerable subsequent retellings, Fuller transformed the chaotic events of 1927 into a conscious, methodical program. Inspired by an exquisite sense of self-importance and a humbling sense of obligation, Fuller made the retrospective decision to conduct his life as a rigorous scientific experiment and to become, in effect, a fully documented living hypothesis. "One of the first things I found myself doing back in 1927 was saying that I think one of the things I ought to start off with is to make a list of everything that I can remember ever happening to a human being."[5] This "list" of 1927 would grow into the 750-volume Dymaxion Chronofile that, supported by correspondence, manuscripts, drawings, blueprints, models, photographs, films, video- and audiotapes, and newspaper clippings (not to mention love letters, overdue library notices, and invoices for an immense number of unpaid bills), resulted in some forty-five tons of compulsive self-documentation. There is nothing like it in American intellectual history, and the implication is inescapable: if it is not in the Chronofile, it most likely did not happen.

The outlines of the story, accordingly, are pretty well known. Richard Buckminster Fuller was born in 1895 into an old New England family and spent his boyhood summers in Maine, immersed in the marine culture of Penobscot Bay. Admiration for the design, fabrication, operation, maintenance, and underlying hydrodynamic principles of boats remained with him throughout his life and often served as a standard against which to judge other, less enlightened industries.[6] Following a private secondary school education and a famously inglorious episode at Harvard—from which he was first dismissed and later expelled "for lack of sustained interest in the processes within the University"—he found himself working in a cotton mill in Québec and then in the meat-packing industry in New York and New Jersey. Fuller would later reassemble these experiences into logical and indeed methodological stages in the formation of a comprehensive worldview. In the mills he gained a "dawning awareness of [the] addition of value (or wealth) by manufacture"; the packing house gave him a perspective on "broad-scale, high-speed, behind-the-scenes human relations in the give and take provisioning of men's essential goods."[7] Even his athletic pursuits at the Milton Academy imparted to him the "intuitive dynamic sense" that he viewed as fundamental to his "anticipatory design science."

Likewise, the importance of his naval service in World War I lay, for this assertively apolitical autodidact, not in contributing, however modestly, to a world made safe for democracy but insofar as his training at Annapolis left him with "an enthusiasm for scientific methodology" and for complex technical systems whose patterns and principles were wedded within "a reciprocating, dynamic totality." With his inimitable gift for compressing a highly articulated *Weltanschauung* into a single sentence, Fuller summed up his wartime experience this way: "I learned the process of conscious self-attunement toward the understanding of principles and their subsequent teleologically translated anticipating effectiveness . . . all of which attuned comprehension of principles invariably was reduced to generalized complex equations by a process of decisive and swift differentiating-out of the problem's complementary functions."[8]

Somewhat more concretely, he gained exposure during the war to radio telephony and naval aviation, and in 1917 Fuller married Anne Hewlett.

Anne was the daughter of James Monroe Hewlett, a prominent New York architect, muralist, and the inventor of a lightweight fiber-concrete construction block that would soon compete with her for Bucky's attention. The first years of their marriage were difficult, punctuated by the devastating loss of their first child to meningitis. Although he experienced intense feelings of guilt at the time, Bucky gradually sublimated his grief into an intellectualized "resentment" at their substandard living conditions and a programmatic mission of "the housing of children, large and small."[9]

In 1922, in the wake of their tragedy, James Hewlett invited his son-in-law to go into business with him as president and principal sales representative of the Stockade Building System. Fuller's business correspondences over the next five years, together with the almost daily letters from Anne written during his prolonged absences, are the source of most of what can be reconstructed of the period leading up to the crisis of 1927. Considering all that was to follow, it is remarkable how utterly unremarkable the record is. There are pages on pages of correspondence with disgruntled customers, anxious partners, and impatient contractors, all of which are executed in a mildly verbose but otherwise businesslike manner. Only the voluminous correspondence from his wife—always addressed to "Darlingest Buckie" and always signed "Your very own Anne"—illuminates his inner life, but even this is notable for its startling lack of reference to the larger world: no mention is made of books read or lectures attended; of visits to concerts or museums; of current developments in science, politics, or the arts. If Spaceship Earth is operating in those years, it is on autopilot.

Although there is nothing particularly visionary in the Stockade system ("It is the latest innovation in wall building"), Fuller committed himself to the business and worked unflaggingly, apparently unaware that his career was about to implode. During this period he and Anne were beset by financial insecurity and strained by his extended absences, and in early 1927 Anne began to experience anxieties over her second pregnancy. From their apartment in New York she wrote repeatedly of her fears that Bucky would have an accident or a breakdown, that he was sleeping too little and drinking too much, and that "we will never be together again, happy again." Nor were their periodic reunions always blissful, as suggested by "that nasty night in Chicago when you staid [*sic*] out so long and then sort of banged me around when you came in."[10]

In July Fuller was abruptly forced out of Stockade in a managerial coup, and a month later their second daughter, Allegra, was born. This was, indisputably, a traumatic period, and a flurry of letters depict a man who felt in turns defeated and defiant. In subsequent reconstructions this confluence of events carried no such ambiguity, however, but became "a critical detonation point" and the defining moment in his life:

> Now in 1927, when our daughter Allegra was born, we had no money, and obviously under those conditions I ought to have gone out to earn a living. But it was just at this moment that the kind of picture I have been describing was looming before me and I didn't see how I could escape doing something about it.

Or again:

> I set about [in 1927] to see what a penniless, unknown human individual with a dependent wife and a newborn child might be able to do effectively on behalf of all humanity.

And again:

> In 1927 I committed myself for the rest of my life to undertaking the solutions of problems which were not being attended to by others which experience taught me would, if effectively solved, greatly advantage society and if left unattended, would bring about comprehensive disadvantage for all.

And again:

> In 1927 I undertook a thirty-year series of experimentations . . . in relation to the individual and his functioning, and in relation to the questions of whether and how he can take the initiative in regard to various challenges.

And once more (this time for emphasis):

> I saw in 1927 that there was nothing to stop me from trying to think about how and why humans are here as passengers aboard this spherical spaceship we call Earth.[11]

All that is missing from this divinely inspired comedy is the *terza rima*, for what Fuller would repeatedly describe is nothing short of a descent into the infernal depths of personal and professional failure followed by a miraculous rise through the purgatory of neglect to the paradisiacal heights of international celebrity. It is, by all accounts, an inspiring and even epic tale, even though it is not supported by the record that he himself so assiduously maintained. In fact, his personal and business correspondence continued throughout this period with no discernible evidence of a philosophical breakthrough or a psychological breakdown. Far from "peel[ing] off from conventional livelihood preoccupations" and embracing his destiny, Bucky mounted an aggressive job search and soon landed a position—at $50 per week plus commissions—as midwestern sales representative for a manufacturer of asbestos floor coverings. His boast that he at no time engaged in any self-promotion is preposterous; it's not clear, in fact, that during this period he did anything else. He did not, in 1927–28, "renounce the use of words" and in fact delivered twenty public lectures during the month of April alone.[12] Bucky would regularly claim that in 1927 he resolved to use himself as a measure of "how much the contemporary individual might be able to effect," but the record provides no evidence that he set any such conscious agenda at the time.

If events were embellished by the subsequent requirements of the experimental subject he named "Guinea Pig B[ucky]," it is nonetheless clear that an epiphany enveloped him during these extraordinary months during which he laid the foundations of the Dymaxion worldview. He did so in a treatise, *4D Time Lock*, that teetered precariously between a highly technical business plan and the ravings of a sidewalk preacher. "*4D* is my own creation," wrote Bucky to one of his financial backers, "or rather has been revealed through me."[13]

III

At the risk of belaboring the obvious: Anne gave birth to Allegra in August; exactly nine months later, following an intense spasm of generative activity, the expectant Bucky brought *4D* to term and proudly delivered what the title page tellingly announced as "some pregnant prognostications." Having held himself responsible for the death of one child, he was determined to ensure the survival of his second daughter, no matter what it took.

Fuller's five-year experience selling the Stockade Building System had turned him into a propagandist of rational building methods, and his departure from the company deepened rather than derailed that commitment. As he reported to his mother in May 1928, after swearing her to the utmost secrecy: "I will send you some time soon a very strong paper that I have been writing for many months . . . and which will astonish you. I know from the secret conferences to date that it is going to make considerable excitement when it is published."[14] A few months later he wrote again, this time begging her to sell off her real estate holdings in Cambridge because "in a year or so when my 4D houses are ready," the value of land will plummet. In one breath he is assuring her that he has "struck a gold-mine" and in the next that his motives are wholly altruistic. "The main thing I want to impress upon your mind is that this whole affair is no 'scheme' of mine, but merely the observance of truths which people overlook in their great rush for survival." One cannot miss the messianic fervor in his voice, nor his faith in the truth he had been appointed to reveal: Echoing the prophet Isaiah ("And it shall come to pass in the end of days . . ."), he flatly declares that "there is no question that what I have predicted will come about."[15]

4D is the product of a gifted but volatile mind in the "throes of mental anguish" and anticipates all of the endearing, and often maddening, eccentricities for which he would later become famous—the sixteen-hour lectures, the preference for monologue over dialogue, the spontaneous neologisms, the startling logical reversals ("Wind doesn't blow; it sucks."). At the most prosaic level, *4D* is a proposal for a new type of small house that Fuller earnestly hoped would be embraced by the building trades, the financial industry, and the architectural profession. More grandly, it is a vastly conceived rant against the organization of modern civilization. Respecting Bucky's own wishes, we must avoid any temptation to separate the two.[16]

At the heart of Fuller's treatise is the "Lightful House," which, rather than being supported by heavy, load-bearing walls, was to be suspended from a central compression mast. The mast serves as a central service unit that delivers heat, water, and electricity to the floors which radiate outward on cantilevered arms; in successive refinements this rather ungainly four-sided, cantilevered structure would evolve steadily toward the six-sided Dymaxion house suspended by cable stays, equipped (rather than "furnished") with standardized and industrially mass-produced "Lightful Products." It was designed to be self-contained, lightweight (three tons, to be exact), and thus easily transportable by air, and, significantly, to be assembled virtually anywhere in one day, "complete in every detail with every living appliance known to mankind built-in." Going beyond Le Corbusier's call for a house that *works* like a machine, Fuller proposed a house that *is* a machine.[17]

4D is not really about houses, however; it is a spiritual meditation on time, the supramaterial fourth dimension of experience and the true measure of industrial society. "Without legislation recognizing it, the world is now on a time standard instead of a gold standard," he wrote.[18] Building a house in six months as opposed to assembling it in a day; armies of architectural draftsmen each designing his own window frame; guilds of tile-setters jiggling one-inch ceramic squares into place when an entire bathroom unit could be stamped out of aluminum—such anachronisms seemed to him to be evidence of what George Bernard Shaw, according to a newspaper clipping that found its way into the Chronofile, had decried as "a terrible time-lag" that afflicted the public imagination. It was a short step from time-lag to Time Lock, "in which the great combination is revealed" (fig. 2.1).[19] All that is missing are initiation rites and a secret handshake.

The design details of the successive Lightful, 4D, and full-blown Dymaxion houses have been closely analyzed, and it is not necessary to do so again.[20] Fuller himself, moreover, sought to deemphasize the artifact: "You must dismiss the idea that we are organizing around a material unit," he wrote to a sympathetic backer. "We are organizing around service, abstract satisfaction, etc. The easiest part will be the fabrication of the house."[21] Fuller clearly saw himself as the pioneer of a new industry, in much the same way that Thomas Edison invented not the lightbulb but the electrification of society and Henry Ford invented not the automobile but the mechanization of individual human transport. Henry Ford once estimated that a Model A would cost $43 million if he were to make just one; the secret of its accessibility is mass-production, a secret that the housing industry—"the most ignorant and most prodigious of men's fumbling activities," as Fuller later put it—seemed determined not to discover. "We have industrialized all the non-essentials and the near essentials," he wrote. "We are at a point when those in charge of capital must realize that we have overlooked the most essential product for industrial production—the home."[22]

Figure 2.1 Cover for *4D Time Lock*, showing a combination lock superimposed on an hourglass, superimposed on a polar projection of the globe, with a gnomon planted on the North Pole. Source: Special Collections, Stanford University Libraries.

The program Fuller was attempting to articulate in this formative period suggests a man out of time. The absence of the requisite technical infrastructure—materials and manufacturing—left many of his potential backers feeling that Bucky was, at best, an ungrounded visionary, while his mystical appeals to "the Cosmic Trinity of Stability" (the theological counterpart to his beloved triangle) and his bizarre iconographic drawings recalled for others the secret guilds of the Middle Ages and the microcosms and macrocosms of the renaissance magi (fig. 2.2).

Just as his 4D houses were designed "from the inside out," so he hoped to begin with the house—"the last primary area of man's activity yet to come importantly under the effect of the industrial equation"—and redesign the whole of society from the inside out. Demographic trends alone make a solution to the housing crisis imperative, he argued, but once it is correctly addressed, Fuller blithely predicted that society's remaining problems "will practically solve themselves." Properly understood not as a structure but as "philosophy . . . mechanically applied," 4D would put an end to the drudgery of housework and "the bunkum of archaic design and assininic aesthetics" but

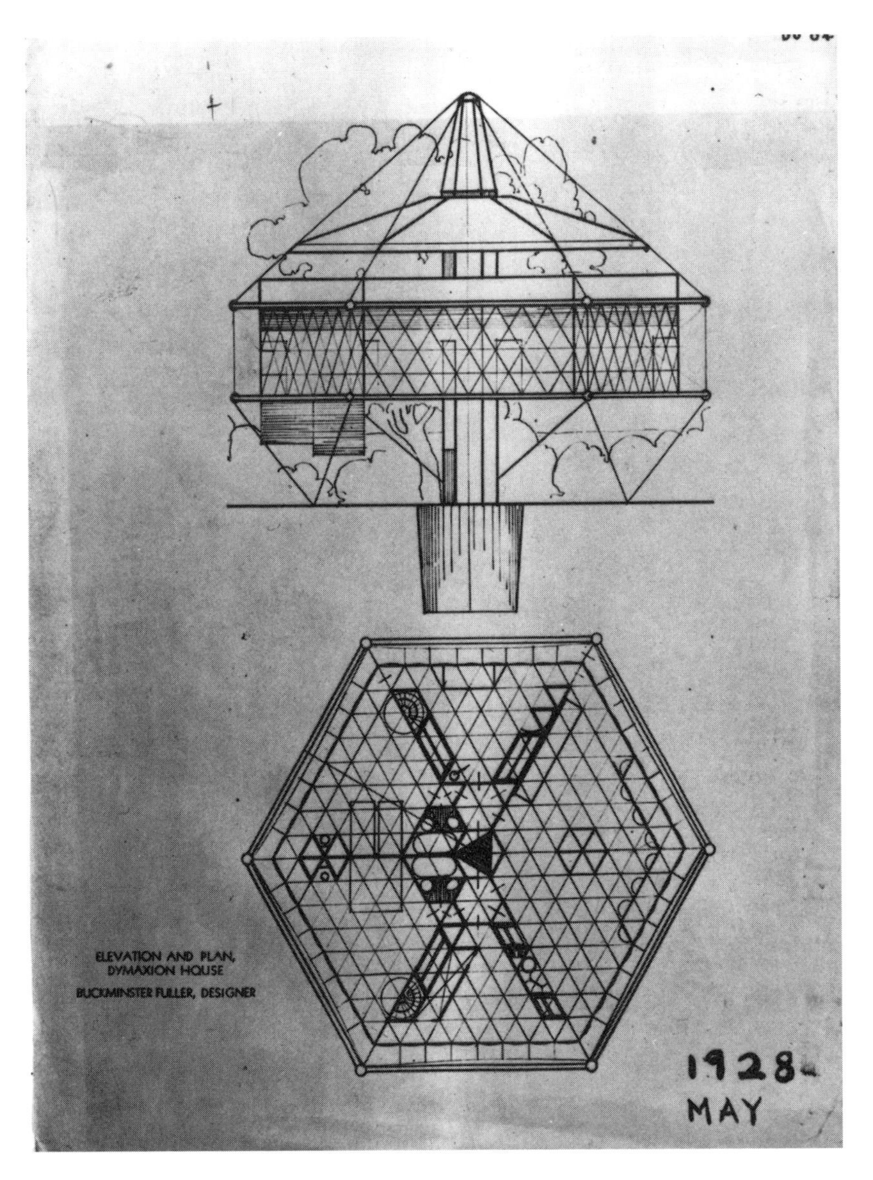

Figure 2.2
"Elevation and Plan, Dymaxion House" (1928). This extraordinary image was published by the Harvard Society for Contemporary Art some months before Fuller would deliver "my first lecture in a real auditorium" (R. Buckminster Fuller to Anne Fuller, March 3, 1930 [Fuller MSS, M1090:2:56:12]). Source: Special Collections, Stanford University Libraries.

also to the derivative social afflictions of unemployment, poverty and crime: "With so much time and such perfect privacy of the most minor member of the family, the minds will be turned to a more philosophical and rhythmical contemplation of life."[23]

Bucky labored feverishly to wrestle the two thousand pages of notes he had accumulated into presentable form and was given the incentive to do so by the American Institute of Architects (AIA), whose annual meeting was held in St. Louis in May 1928. Before a small group of eighteen attendees he alluded to Lindbergh's conquest of the spatial universe and called for "a new spirit of individualism, delivered of its last material impediment of self-consciousness, [to] fly forth from St. Louis once more, this time, to girdle the temporal universe."[24] Thus would the prowess of industrial technology be brought to bear on the problem of housing. His reception, by Fuller's own estimation, was as rapturous as that given to Lindbergh the year before; as he reported to a friend, "The remarks on the philosophy, literary value, etc. as well as the subject matter were enough to arouse the old ego were it not for the frightful seriousness of it all."[25] The official proceedings of the meeting record a rather different view, however. The president of the AIA himself felt moved to write a resolution "Against All Standardization," which sought to dispose once and for all of the specter of the factory-built house. Anticipating the arguments of the Critical Regionalists in the 1980s and the anti-Globalists of the 1990s, the resolution lamented that "local characteristics are fast disappearing in this era of common thought and mechanical advancement. Communities are coming to look more and more like peas of one pod, and a certain commercialism is making itself more and more evident in the type of architecture universally enjoyed throughout the country."[26]

This rebuke was enough to turn the AIA, in his eyes, from "a selected group of the country's leading architects" into a self-appointed clique of "feudalistic" potentates. Other potential allies did not fare much better. After workers in Donald Deskey's office accidentally rammed a sofa into a model of the Dymaxion house (which had been unceremoniously relegated to a storage room), the Industrial Designers were summarily demoted from prophets of the new age to a mercenary cabal of style-driven hucksters and charlatans. Fuller distanced himself from the financial community, which became little more than a scheming plutocracy, and dismissed the American Technocracy movement as the idle daydream of a bunch of unemployed engineers.[27]

Unfazed by these mounting indignities, Fuller had his opus printed and bound and mailed it out to "some 200 men and women of high public esteem." His detailed logs confirm that the vast majority of them did not respond at all, including—doubtless to his great astonishment—such intellectual soul mates as Bertrand Russell, Albert Einstein, and Henry Ford. Some at least acknowledged receiving it, including the president of Harvard University, to whom Fuller presented himself as a fifth-generation Harvard man; it was nonetheless returned the same day with the apology that "President Lowell is far too busy at this time of the year to give it the attention it may deserve."[28] Others thanked him and promised a detailed response "at the earliest possible moment," and a few—including his increasingly alarmed father-in-law—seem to have made an honest effort to get through it: "Possibly there is something in Mr. Fuller's idea," wrote a prominent advertising executive whom Bucky had hoped to enlist to the cause, "but it is so well concealed in his language that I have not discovered it."[29]

During this period Fuller assured his friends and relatives that he was only a few weeks, a few months, or at most a few years from single-handedly completing the final stage of the industrial revolution. His obstinacy is to his credit and may, in fact, have been the trait that saw him through this period of precarious mental instability. From ranting to his father-in-law about machinery as "the one and only Messiah's gift . . . for the salvation of the spirit in mankind" to his unreal assurance to a correspondent that "you may count [this] as the greatest letter which you will have ever received," to his unsettling boast that "the time-relativity discussion was quite equal to Einstein's," it is apparent that his personal situation was fragile in the extreme.[30] In the face of public indifference and professional repudiation, he stubbornly maintained that "those who have read deeply into *4D* have found their appreciation of life and its progression vastly improved, and in the void a new, fuller and happier sense of mental poise and purposefulness than they have ever had before."[31] Obviously, Buckminster Fuller is speaking here of his own aspirations: the "*fuller*" sense of mental poise he longs for, together with his frequent warnings about the devastating effects of "*buck*-passing which has, through each age, selfishly retarded progress," suggest the depths of his personal crisis.

Conclusion

By 1930 Bucky had regained his own sense of "mental poise and purposefulness," though he remained eccentric and iconoclastic to the end.[32] By this time there were growing numbers of inquiries, lecture invitations, press notices, and a widening circle of friends, and it is clear that he was beginning to gain a hearing. But the long shadow of 1927, Bucky's annus mirabilis, hung over him: it reveals itself in the unresolved tension between mechanical idea and social reality that would be at one and the same time the hallmark of his genius and the limiting factor to it.

There is no issue to which Bucky was more sensitive than that he had somehow "failed"—as if this charge rekindled the feelings of self-doubt that were burned into his psyche in 1927.[33] Thus his constant recitation of his accomplishments and his continual reimagining of his story. But if he was not a failure, neither was he a success by any reasonable measure—or even by his own unreasonable one! Bucky's lifelong campaign was not to invent a new kind of house, car, or map. It was to use his "anticipatory design science" to complete what might be called "the unfinished project of industrialism," and this he manifestly did not do.[34] It is not enough to point to his collection of honorary doctorates or to three hundred thousand geodesic domes scattered about the globe, from the northern Greenland of NATO to the Moscow of Khrushchev and Nixon to the fairgrounds of Montreal's Expo 67 to the communes hidden away in the Santa Cruz mountains. An honest reckoning would have us ask why there are no domed communities in downtown Detroit or suburban Chicago; why Americans still drive to their neighborhood grocery stores in modified military assault vehicles rather than three-wheeled Dymaxion teardrops; and why "Synergetics," rather than such parochial disciplines as mechanical engineering and economics, is not taught in our universities.

Bucky did not suffer from a failure of vision—there is scarcely a more visionary figure to be found in the annals of twentieth-century thought. But the gap between idea and reality that opened up in 1927—and the colossal demands of what his friend and admirer Hugh Kenner delicately referred to as "Bucky's highly mobilized ego"[35]—prevented him from making the concessions that might have permitted the best of his ideas to gain traction. Bucky boasted that he had "a genius for getting into trouble and then getting out of trouble when I had been displaced and moved into an area where there was something I could get hold of, like a piece of machinery. *Anything you could weigh or feel or apply yourself to was fine, but not the dealings with the patterns of arbitrary customs.*"[36] This sounds self-servingly romantic, and it is true that literature has always favored the lone hero who stands up to custom and convention. But Bucky's lifelong preference for the clean objectivity of machinery over the "arbitrariness" of human behavior must also be seen as the incapacity that prevented the revelations of 1927 from moving from a quasi-religious epiphany to an actionable plan. Kenner, attentive to the generative metaphor that runs through Fuller's work, refers to the Dymaxion plan as "the stillbirth of an industry."[37]

Too much of the secondary literature—one is tempted to say *all* of it—reads backward from the astonishing creativity and preternatural determination that marked Bucky's lengthy career, and he most assuredly did so himself. Reading "forward," however, one finds that the picture does not resolve itself so easily into tidy little hexagons. The voluminous documentation of the Chronofile simply does not support Fuller's oft-repeated claim that in 1927 he conceived a deliberate plan to conduct his life as an experiment in "how much the contemporary individual might be able to effect." To the contrary, the record suggests a period of capriciousness, confusion, and calamity. It was also, however, a period of indisputably great creativity that opened onto all that would follow. He would live the rest of his extraordinary life, like one of his own tensegrity structures, suspended between these two poles.

3 R. Buckminster Fuller

America's Last Genuine Utopian?

Howard P. Segal

This essay will explore six key aspects of R. Buckminster Fuller's life and work and place them in broad historical context in order to appreciate more fully his contributions, as well as his limitations. Too often Fuller's admirers and critics alike have treated him in a vacuum, detached from developments before and after his time. Any connections drawn with earlier visionaries, much less with those of our day, have generally been superficial.

R. Buckminster Fuller was above all else a technological utopian, someone who believed that technological advances would vastly improve—if not literally perfect—society. Far from being the first such American technological utopian, he was part of a long and rich tradition whose origins derive from European ideas and visionaries. Thus the common characterization of Fuller as a quintessential American dreamer or inventor must be at once revised and expanded. We must also reexamine the frequent emphasis on Fuller as primarily a twentieth-century successor of Margaret Fuller (he was her grandnephew) and other nineteenth-century American transcendentalists (fig. 3.1).

Utopias

Utopia means the perfect society. The term was coined by Thomas More (1478–1535), lord chancellor of England, as epitomized in his *Utopia* (1516). Previously, as in Plato's *Republic*, there was no expectation of genuine improvement but, instead, a lamentation of the insurmountable gap between the real and ideal worlds. "Platonic forms" like the *Republic* reflected that avowedly human condition. More, however, was the first to hold out the prospect, albeit dim, of actually establishing a perfect society and thereby altering human nature. It is true that the term *utopia*, which More coined from Greek roots, means "nowhere" (although it also means "good place") and that More considered human nature depraved. Nevertheless, he considered utopia a possibility, so that *utopian* need not mean "forlorn." From More as well came a utopian tradition eventually traced back to Plato, a tradition that includes Fuller.

Figure 3.1
Margaret Fuller (1810–50), R. Buckminster Fuller's great aunt.
Source: Special Collections, Stanford University Libraries.

As more than thought experiments—as visions that might actually be realized—utopianism is rooted in two European developments. The first was the growth, starting in the sixteenth century, of the Enlightenment faith in the power of reason to achieve steady human improvement, the logical outcome of which was unprecedented belief in the prospects for utopia. The second development was the rapid technological advances that, beginning in England's textile industry in the mid-eighteenth century, led to the English industrial revolution and its spread to most of the rest of the world. These developments are intertwined, for it was the Enlightenment viewpoint, combined with those enormous technological advances, that made utopia look like an achievable goal. Most utopian schemes over the centuries have presupposed the availability of adequate food, clothing, and shelter, but only in modern times could the availability of such essentials be taken for granted.

Indeed, it was only in late nineteenth-century and early twentieth-century America and parts of Europe, when Fuller was growing up, that the long-standing assumption that natural resources were finite gradually gave way to the belief that, thanks to numerous technological advances, natural resources were infinite. Tools and machines to extract, transport, assemble, and process raw materials and to turn them into finished products were the key developments.

The European utopians who forged the first real connections between utopianism and fulfillment through technology were varied. Their initial ranks included the Pansophists, a small number of late sixteenth-century and early seventeenth-century visionaries, including Francis Bacon (1561–1626), the scientist, philosopher, and, like More, lord chancellor of England; and the French philosophers and social critics Marquis de Condorcet (1743–94), Henri de Saint-Simon (1760–1825), Auguste Comte (1798–1857), and Charles Fourier (1772–1837). Later visionaries included the British industrialist and communitarian Robert Owen (1771–1858) and the communists Karl Marx (1818–83) and Friedrich Engels (1820–95).

Yet for all of their commitment to the power of technology, none endorsed unadulterated technological advance. That is, they did not see technology as the royal road to utopia. All insisted no less on other values. For example, Fourier insisted that there could be no utopia unless individuals could fulfill their varying sexual and other pleasure-providing desires. Meanwhile Marx and Engels insisted on a variety of work and leisure activities to avoid exhaustion and boredom in the highly mechanized future they envisioned.

Fuller as Utopian

Fuller has been repeatedly criticized for an excessive focus on technological advances, to the neglect of other aspects of a more fulfilling existence. By definition as a technological utopian, Fuller was a technological determinist—that is, a believer in the notion that technology shapes society, economy, and culture. Although historians and other scholars of technology now know otherwise,[1] Fuller was very much in line with countless other Americans of his day in assuming that, with the proper technological achievements, improvements in other domains would invariably come about in due course. Given the enormous changes in transportation, communications, commerce, construction, and other areas that ordinary Americans witnessed in everyday life in the late nineteenth and early twentieth centuries, it is hardly surprising that Fuller, like some other Americans, went a step further and envisioned technology as a panacea for nearly all human ills. For him it was a natural progression, not a deviation.

This does not, however, deny Fuller's concern with the human element in technology. Unlike many other technological utopians, Fuller was also interested in technology insofar as it would enhance the quality of life of ordinary people. He had an ethical dimension all too rare among technologically oriented visionaries.

Moreover, notwithstanding his abundant faith in technology, Fuller understood in young adulthood that, contrary to prevailing sentiments of his day, the earth's resources remained finite. To his considerable credit, and long before it became conventional wisdom, he wisely did not equate humankind's growing power over nature with an endless supply of raw materials. In recognizing what was later termed the "limits to growth," Fuller was turning back to the outlook shared by most Americans throughout the eighteenth and early nineteenth centuries, prior to America's own industrial revolution.[2] Fuller's inventions accommodated this reality, unlike some of his contemporaries, such as Herman Kahn and Julian Simon, who assumed that inventive humans either would always discover natural resources in hitherto unexplored places or would readily create adequate substitutes.

If Fuller was part of that historic tradition of technological utopianism, he was simultaneously the last popular American utopian who engaged in utopian thinking, speaking, writing, and building for their own sake rather than, as is common today, for commercial reasons. Fuller endured decades of poverty, indifference, and ridicule before obtaining respect, influence, income, and a following.

Fuller was not primarily concerned with earning a living—much less becoming rich—from his visionary activities. He was primarily concerned with trying to improve the world. Even when, in the 1920s, Fuller had to support his wife and child, he never indulged in speculation and invention just to make money. Far from it: in 1927, at age thirty-two, and after repeated business failures, he renounced personal success and financial gain. As he put it in his *Grunch of Giants* (1983), "I learned very early and painfully that you have to decide at the outset whether you are trying to make money or to make sense, as they are mutually exclusive."[3] Given the coming Great Depression, maybe it was just as well. Yet this altruistic stance might also have been self-serving, a convenient rationale for Fuller's business misadventures.

By contrast, most recent prominent American utopians—such as Alvin and Heidi Toffler, John Naisbitt and Patricia Aburdene, Michael Dertouzos, Nicholas Negroponte, Bill Gates, and Virginia Postrel—have helped to transform forecasting into a lucrative big business. They write and speak largely for the marketplace, including the promotion of their respective business or academic or other enterprises. And they readily revise and update their predictions when, as is commonplace, events undermine their forecasts. Rarely, if ever, do they admit their errors.

To be sure, Fuller also resisted conceding mistakes and was also not above making up numbers, so one cannot exempt him from similar criticisms. But unlike the foremost visionaries, dating back to More and including Fuller, these contemporary prophets provide no genuine moral critique of the present, no serious effort to alter society for higher purposes. There are a few exceptions, especially Gates's admirable philanthropy—as epitomized by his being chosen, along with his wife, Melinda, and rock star Bono, as *Time* magazine's 2005 "People of the Year." But the rest are not the kind of socially responsible citizens that Fuller and most of his predecessors were.[4]

This is not to deny Fuller's acceptance of relatively large speaking fees in his later years, when he gave hundreds of lectures annually. Still, Fuller's life and work had an integrity that should permanently enhance his reputation. This will be the case no matter what the long-term significance and influence of his material and professional achievements. Moreover, his earnestness is definitely shared with Margaret Fuller, Ralph Waldo Emerson, and others of their day. Yet the transcendentalists did not take a vow of poverty, and one can only wonder if Fuller would have renounced "making money" versus "making sense" if, by 1927, he had been financially successful to any degree. The real issue, though, is not his or any other visionary's wealth or poverty but rather the extent to which one's analyses and predictions are affected by economic self-interest.

Curiously, those contemporary high-tech zealots from the Tofflers onward rarely, if ever, consider the prospect that, far from being original, their respective crusades are only the latest in that rich Western tradition of technological utopianism outlined above. Because of their historical ignorance, they can naively proclaim the alleged uniqueness of their particular technological advances, above all the extent and the speed of the transformations to be brought about by technological advancement. They thereby skip over earlier claims about the alleged grand powers of earlier technologies—and not only with the English and American industrial revolutions but also with that of the Middle Ages.[5] Fuller was hardly a professional historian, but he did grasp that he was not the first technological utopian—though surely the best.

Moreover, probably none of the contemporary high-tech prophets have any knowledge of how old-fashioned they really are in their celebrations of technology's prospects for transforming the nation and, in due course, the world. As they insist, their particular advances may indeed operate and spread with unprecedented speed but not necessarily with unprecedented transforming powers—as compared, for instance, with life in rural England during its traumatic industrial revolution of the eighteenth and nineteenth centuries.

Fuller escapes much of this criticism because he grew up and began working in a cultural climate, at least within the United States, still fundamentally hopeful about the future and about technology's transforming prospects. But Fuller's later years and growing fame were certainly tarnished by his failure to recognize that the world had changed, as evidenced by questions about his ready acceptance of the Pentagon's use of giant helicopters in the Vietnam War—a war he didn't endorse—because they might later be employed to transport and deposit ever-larger geodesic domes elsewhere in the world and for other nonmilitary purposes.

Fuller was the first major American utopian who argued that the realization of utopia was possible within our own lifetimes rather than, as with all earlier utopians, either at least two generations or more ahead or virtually impossible. Fuller's *Utopia or Oblivion: The Prospects for Humanity* (1969) makes this point explicitly:

> Not only did all the attempts to establish Utopias occur prematurely (in [the absence of our contemporary] technological capability to establish and maintain any bacteria- and virus-immune, hungerless, travel-anywhere Utopias), but all of the would-be Utopians disdained all the early manifestations of industrialization as "unnatural, stereotyped, and obnoxiously sterile." The would-be Utopians, therefore, attempted only metaphysical and ideological transformations of man's nature—unwitting any possible alternatives. It was then unthinkable that there might soon develop a full capability to satisfactorily transform the physical energy events and materials of the environment—not by altering man, but by helping him to become literate and to use his innate cerebral capabilities, and thereby to at least achieve man's physical survival at a Utopianly successful level. All the attempts to establish Utopias were not only premature and misconceived, but they were also exclusive. Small groups of humanity withdrew from and forsook the welfare of the balance of humanity. Utopia must be, inherently, for all or none. . . .
>
> . . . This moment of realization that it soon must be Utopia or Oblivion coincides exactly with the discovery by man that for the first time in history Utopia is, at least, physically possible of human attainment.[6]

The classic example of a utopia allegedly *not* intended to be established is Plato's *Republic.* The starting point for utopias that *could* be established is More's *Utopia.* For centuries thereafter, and continuing at least as late as James Hilton's *Lost Horizon* (1933), utopia was usually discovered by Western travelers who came upon it by accident, such as through erroneous maps, storms at sea, or airplane crashes or, as with Edward Bellamy's *Looking Backward: 2000–1887* (1888), through falling asleep and awakening in utopia. Conditions that eventually brought utopia about included wars; postwar peacetime negotiations;

natural disasters; and clashes between continents, nations, classes, races, and even sexes. These utopias were usually placed in the contemporary time of their authors. But as more of the world was explored and known, it became increasingly necessary to place utopia in unexplored, exotic places in order to claim some originality—for example, under the sea, inside the earth, or in outer space, as with the writings of Jules Verne, among others. As, however, these sites in turn became explored and relatively familiar, it became necessary to project utopia into the future.

At first, beginning with More, there were vague expectations of utopian fulfillment in the distant future—for example, the works of the various European visionaries just discussed—but usually no particular dates. Eventually, though, there arose visions to come about in a specified time within reach of the next generation or two, as with *Looking Backward*; and, later, within one's own lifetime or that of one's children—as with the date of 1960 in the landmark World of Tomorrow exhibit at the 1939–40 New York World's Fair. With Fuller's *Utopia or Oblivion*, however, came the elimination of any delay: *the future was now*. This outlook has had enormous influence on later visionaries, from the Tofflers and Naisbitt and Aburdene through Gates and Dertouzos, regardless of their acknowledgment of Fuller's work.

The extent to which this belief on Fuller's part was belated or lifelong cannot be determined conclusively, but it surely was rooted in developments dating to his childhood and young adulthood. For example, he had endured personal and physical setbacks long before his 1927 decision not to commit suicide while standing on the shores of Lake Michigan: his terrible vision even as a toddler, the loss of his father when he was an adolescent, the taunts by his sister Leslie, and the leg injury that prevented him from playing football in college. Moreover, that decision itself may have reinforced his preexisting practical bent, his dedication to a "higher purpose" than financial success. Such a "can-do" spirit could in turn have spurred his "future is now" stance once the technology was available to transform the world.

From Idea to Reality

More than any other American utopian, Fuller bridged the common gap between those who write seriously and significantly about utopia and those who attempt to build it. Where such prominent European utopians as Owen and Fourier at once wrote about and constructed or inspired actual communities, American communitarians have commonly been bereft of intellectually substantive plans; and most genuinely intellectual American visionaries—not least, those nineteenth-century transcendentalists among whom Margaret Fuller was a leading figure—have usually lacked the practical skills to establish successful communities. R. Buckminster Fuller is a notable, *if only partial*, exception.

By way of background, utopianism in general has taken various forms over the centuries, beginning with prophecies, speeches, and writings; continuing with political movements, actual communities, and world's fairs; and moving recently to virtual communities in cyberspace. In late nineteenth-century and early twentieth-century America and parts of Europe it became clear that only thoughtful, reflective tracts could respond convincingly and appealingly to the complex contemporary developments—for example, industrialization, urbanization, immigration, poverty, and class conflict—that had by now turned actual utopian communities into marginal phenomena. This included such relatively successful American communities as the Shakers, Oneida, and Amana that combined traditional agriculture with industry and manufacturing for greater profits, balance, stability, and longevity. The key figure credited with the establishment of American utopian writings as a serious, modern intellectual genre was Bellamy (1850–98), whose extraordinarily popular *Looking Backward* demonstrated that writing was a superior means of examining these contemporary problems and of providing richer alternatives to existing society than even the foremost utopian communities. Admittedly, utopian speculation was obviously less risky when placed on paper than when placed on land, though Bellamy himself helped to launch a political crusade.

To be sure, utopian communities in the United States and elsewhere have continued to be established through the present. There have been periods of extensive community building, most notably the counterculture of the late 1960s and early 1970s and, more recently, cyberspace communities. But by the mid-twentieth century, no American utopian save Fuller was capable of both writing and building, of both theory and practice.

True, Fuller did not establish actual communities. But he invented enough actual components of potential communities to distinguish himself from virtually all other American utopians: the Dymaxion car, the Dymaxion house and bathroom, the Dymaxion air-ocean world map, and geodesic dome. In all, Fuller received patents for twenty-eight inventions, a notable achievement in itself. Moreover, Fuller proposed land-based cities covered by geodesic domes, tetrahedronal cities floating on the sea, and cloud-structure spheres floating in air. Fuller was a de facto architect long before he got a license, which he was awarded honorarily when he was in his late sixties.[7]

Many prior communitarians had been no more specific than Fuller in their plans, so his providing a mere scaffolding for utopian communities rather than detailed blueprints hardly undermines his legitimacy here. In fact, by designing artifacts that could be both moved and replicated, Fuller readily met a principal challenge facing most earlier (and later) communitarians: how to promote one's vision beyond its base camp, so to speak. If, to take a notable exception, the Shakers met this challenge by building ever more communities, they simultaneously undermined their impact on the rest of America by ruling out procreation by existing members and by instead concentrating on luring outsiders to their ranks. Few other utopian communities created even a second site.

In addition, Fuller's work was in the public eye in two major world's fairs. At the 1933 Chicago fair he demonstrated his three-wheeled, rear engine, streamlined Dymaxion car (fig. 3.2). Seating eleven passengers, and getting remarkable gas mileage, the vehicle performed fine as a prototype. But production ceased because of bad publicity following a fatal accident just as potential investors watched. An investigation later concluded that the vehicle was not at fault.

More directly, and much more positively, Fuller's geodesic dome was chosen for the United States Pavilion at the 1967 Montreal fair. The twenty-story dome has remained structurally intact despite a major fire in 1976. Like so many others involved in world's fairs, Fuller believed that bringing people from around the world to one locale and introducing them to cutting-edge technolo-

Figure 3.2
Dymaxion car.
 Source: Special Collections, Stanford University Libraries.

gies and products would somehow promote world peace and understanding, not just international commerce, and would in turn promote utopian community building. Here, as with his other designs noted above, Fuller followed many other utopian visionaries in taking for granted, rather naively, that new and presumably comfortable material settings would translate into permanent contentment for all inhabitants.

A similarly romantic notion colors, for example, Gerard O'Neill's *The High Frontier: Human Colonies in Space* (1982) and his *2081: A Hopeful View of the Human Future* (1981). The former book convincingly shows how space colonies could be established in the very near future and how life might be carried on within them. The colonies appear to be pleasant, fairly free, quite large suburbs circling the earth. If anything, the society portrayed here—and elaborated on in the latter book—is so harmonious and so earnest that it seems a bit dull, in the manner of *Looking Backward*. One wonders if the inhabitants of O'Neill's colonies wouldn't find life boring and confining once the thrill of being in space had faded—much like the inhabitants of Fuller's proposed land-based cities covered by geodesic domes, tetrahedronal cities floating on the sea, and cloud-structure spheres floating in air.[8]

In all of these cases, however, Fuller can be faulted for not devoting sufficient thought—and design—to the ways in which communities might actually be established and maintained within his various giant scaffolds. The fact that the contents of his 1967 Montreal dome were wholly in others' hands does not negate this point. Ironically, Fuller's schemes appear to be more practical for individual nuclear families than for masses of families or extended families. By contrast, the Shakers, for example, devised innovative means of creating and preserving communities—such as their famous dances—despite their commitment to celibacy.

In the final two decades of his life, Fuller's growing popularity through writings and lectures made utopianism popular again and, to a considerable extent, made it respectable again, in both the United States and much of the rest of the world.

Admittedly, objections to utopianism have been raised since Plato's *Republic*. Utopianism has been persistently criticized as impractical, immoral, deviant, conformist, revolutionary, reactionary, stagnant, authoritarian, and libertarian. Beyond the obvious question of whether specific utopian schemes could ever be implemented has come the traditional concern over forcing individuals, groups, and entire societies to adopt values, institutions, and ways of life that many might otherwise reject—if offered a choice, given the equally persistent association of utopianism with the lack of choice.

Fair or not, this criticism has sometimes been leveled against Fuller as well for his proposed land-based, sea-based, and air-based cities. His contemporary Lewis Mumford was particularly scathing here. Mumford characterized all three of these projects as huge tombs akin to the Egyptian pyramids.[9] How much conformity would be required for these huge communities? And how much individuality would be allowed? For that matter, what institutional arrangements would continue to make life interesting once the thrill of being on land or at sea or even in air wore off?

However, for roughly a century now, there have been two other main objections. First, insofar as the realization of utopia presumes human perfectibility, it is impractical. The technologically assisted horrors of the years since World War I have rendered forlorn any hope of improvement in human behavior. Second, and more perverse, the closer that science and technology bring us to being able to realize utopia, the less desirable it appears. Given the human propensity for selfish and exploitative behavior, achievement of the kind of planning and control required by utopia would result in "dystopia," or antiutopia. The visions of Eugene Zamyatin's *We* (1920) and of the more celebrated Aldous Huxley's *Brave New World* (1932) and Orwell's *1984* (1949) have transformed utopia from something to be yearned for to something to be dreaded.

Fuller's eventual popularity, especially with youth, coincided not only with the growth of the counterculture—however hostile to modern technology many of his young admirers otherwise claimed to be—but also with their and others' desire for more positive visions of the future to consider. Although the cold war would not end until several years after his death, Fuller's avowed optimism offered alternatives to those classic dystopias then still dominant in the public consciousness.

Perhaps the most significant adoption of Fuller by the counterculture was Drop City, a community established in the hills of southern Colorado in the late 1960s. "Singled out by the media as exemplary, Drop City was known for its dome-style of architecture, which combined the principles and methods of . . . Fuller with inexpensive, found materials, such as sheet metal hacked off of junked car roofs." Drop City's founder, one Peter Douthit (a.k.a. "Peter Rabbit"), actually corresponded with Fuller. Where he and three other original settlers—students from the University of Colorado and the University of Kansas—wanted a cheap and quiet place in which to pursue their interests, the media's relentless coverage and its search for weirdness among the inhabitants at once prompted those four to depart and to attract others. But the newcomers were usually transient hippies, and Drop City was eventually vacated. By now most of the property has been developed commercially. The last of the "iconic domes" was removed in the late 1990s.[10]

No less important, Fuller's emphasis on "Spaceship Earth"—a term he coined—and on its fragile ecology contributed significantly to the growing awareness of a world of integrated parts and people, an earth whose environmental systems required constant care. The rhetoric of globalization and of ecology that have become buzzwords in recent years owe a good deal to Fuller's earlier, and more serious, promotion of these basic ideas. Meanwhile, Fuller's very ability to travel around the globe (122 times in all) and his passion for doing so reinforced people's consciousness and mindfulness of the earth's integrity of parts and people and of its simultaneous ecological fragility.

Ironically, Fuller's persistent faith in being able to invent, manufacture, and sell cars, bathrooms, houses, domes—if not yet actual communities—that would supposedly overcome human ills and frailties may have been rendered obsolete by the ongoing genetic engineering advances of the years since his death. The prospect of being able to create human beings with the "desired" physical, mental, and even character traits clearly offers a solution to that fundamental dilemma confronting all prior utopians dating back to Plato and More.

Finally, despite Fuller's popularization of utopianism in the third quarter of the twentieth century, and notwithstanding the appeal of the contemporary visionaries cited above, technological utopianism has steadily lost much of its historic bedrock support in the United States. Consequently, Fuller's legacy is by no means assured.

When my *Technological Utopianism in American Culture* appeared in 1985, two years after Fuller's death, I believed that this phenomenon was already fading in the United States—where it had reached its peak from roughly 1920 until 1980. Technological utopianism had already long faded in Europe. The skepticism toward "progress," and toward progress as technological progress, had pervaded Europe as an aftermath of World War I—the legacy of airplanes, machine guns, poison gas, tanks, and, above all, trench warfare on European soil. By 1985, Americans' hitherto bedrock faith in that same equation of progress with technological progress had belatedly begun to decline. Just as Europeans' prewar Victorian confidence and complacency were shattered by these technological developments, so, in a much slower and still ongoing manner, Americans' historically optimistic outlook was undermined by (1) endless technology-related environmental crises; (2) repeated disappointments over nuclear power and other alleged technological panaceas; and (3) distrust of both public officials and technical experts, a distrust that grew out of the Vietnam War and the Watergate scandal of the Richard Nixon presidency.

Mumford's dismissal in 1975 of Fuller as "that interminable tape recorder of 'salvation by technology'" may have been excessively harsh.[11] But Mumford's worldview was radically different from Fuller's.[12] It reflected the growing disillusionment with technological utopianism.

Admittedly, this analysis of contemporary attitudes toward technology may seem off-base. According to countless newspapers and magazines, television and radio programs, opinion surveys, and, not least, Web sites and Internet discussions, much of the world has for years now been experiencing unprecedented "technomania." *High tech* is the buzzword that encompasses the various advances allegedly justifying this enthusiasm: broadly conceived as virtually all post–World War II technological developments but especially information technology—particularly computers, satellites, the Internet, and the World Wide Web—and biotechnology. Although Fuller died before several of these advances came about—and just two years after IBM introduced the first commercial personal computers—he might still be accorded heroic stature today.

Yet far from technomania's being unique to the 1990s and early 2000s, the historical record demonstrates that most ordinary Americans today are actually much more ambivalent about technology and progress than ever before. They no longer subscribe to the tagline of the old General Electric advertisements that "Progress is our most important product." Besides the obvious examples of nuclear power, oil drilling, and space exploration are concerns such as cell phones reducing drivers' attention and intruding into pristine wilderness, e-mail reducing proper grammatical usage, and the supposedly paperless office generating ever more paper. No less significant, for instance, is the question of prolonging life in modern hospitals by various mechanical means that often undermine the quality and comfort of patients' final days. For that matter, the highly touted unprecedented access to information—the glorious "Information Age" or "Knowledge Society"—has hardly led to unprecedented wisdom, contrary to countless predictions.

In fact, since the 1960s there has been a diminishing faith in experts as supposedly objective analysts rather than as increasingly paid partisans of any cause that will hire them. The utopian claims from the 1960s and 1970s of systems "experts" like Simon Ramo to be able to solve *all* problems through the employment of sufficient experts ring hollow today in light of the cosmic failures of Secretary of Defense Robert McNamara and the other "Best and Brightest" government leaders in conducting the Vietnam War.[13] Countless other examples from both military and civilian realms could be added. In many quarters a healthy skepticism about unadulterated technological advance might actually represent another form of progress. Fuller was certainly not a dull conventional technical expert—and so has escaped some of this criticism—but he was, after all, lauded as an expert visionary and community builder, as detailed above.

No less important, achievement in the twentieth century of so many historic dreams of scientific and technological achievement—be it automobiles, airplanes, nuclear power plants, and, above all, moon landings—invariably brought mixed blessings to most of their inventors, manufacturers, advertisers, and consumers. Nothing is more indicative of the fading of technological utopian fantasies among ordinary Americans and many other people than the relatively muted response on the twenty-fifth anniversary of the first moon landing. In 1994 there was hardly the euphoria that had characterized similar major anniversary celebrations of, say, New York City's Brooklyn Bridge, the first transcontinental railway at Promontory Point, Utah, the first transcontinental telephone line, and the first Ford Motor Company Model T automobile. By 1994 it had become painfully clear to most people everywhere that, contrary to centuries of utopian dreams, the moon landings did not change the world. Like utopian communities, utopian writings, and world's fairs, moon landings could not bring about lasting international—or even national—peace. Nor, as a result, has there persisted in recent decades the uncritical zeal for further space exploration and possible space colonies that was commonplace before 1969. Instead, one finds grassroots skepticism about the value of these megaprojects in view of more pressing needs on Earth. The same skepticism, of course, applies to other megaprojects like the "Star Wars" antimissile defense system and the now abandoned Texas Super Collider.[14] Had he lived longer, Fuller's own megaprojects would not have escaped this revisionist thinking.

When we add to these examples the enormous debates to date on just the early stages of genetic engineering—such as the cloning of certain animals, including pets, and the ability to predict whether terrible diseases will afflict some family members versus others—it is obvious that, more than any other contemporary technological development, this will hardly be deemed everyone's panacea. Decades of heated debates on abortion will pale in comparison to those about genetic engineering.

But Fuller was not, of course, part of these developments. As noted earlier, he was very much in line with countless other Americans of his day in assuming that, with the proper technological achievements, improvements in other domains—not least, human behavior—would invariably come about in due course. His focus on technology as a panacea for nearly all human ills was in a sense part of the end of the so-called Enlightenment project, which had begun in the eighteenth and nineteenth centuries with such efforts as alleviating poverty, criminality, and insanity through new or redesigned asylums. It is hardly a stretch to consider Fuller's—and O'Neill's—projects as the other end of the same spectrum of optimistic rational planning as those asylums.

There is yet another limitation to Fuller's legacy. Whereas during Fuller's lifetime technology was commonly conceived as the solution to widely acknowledged large-scale problems like poverty, irrigation, electric power, and education, in recent decades technology has increasingly come to be viewed as the fulfillment on a small scale of individual needs and desires, ranging from online college degrees to virtual travel to cyberspace relationships. High tech enhances individual sanctity and provides choices in an ever more digitized, programmable world. If one major aspect of high tech is the shrinkage of the size of machines exactly as their power increases—thereby reversing the tradition in which Fuller lived of bigger being more powerful—another major aspect is the consumer mentality that asks what a specific item will do for oneself rather than for society. At this writing, Hewlett-Packard's advertising slogan is "Everything Is Possible." What could be clearer? Similarly, the tagline for most Microsoft advertisements in the same venues is "Your potential. Our passion." There are no qualifications here either.

Thus we see a complete reversal from President John Kennedy's famous 1961 inaugural address—when he suggested that individuals ask not what their country could do for them but what they could do for their country—to asking what one's technology, and implicitly one's country, can do for oneself. Surely Fuller subscribed to Kennedy's position. Indeed, he went far beyond it in seeking such dedication and service on a global, not merely national, scale. He wanted people with integrity to contribute to "Spaceship Earth" rather than to the United States alone. His pioneering use of "globalization" meant infinitely more than the economic bottom line, unlike the early twenty-first century. No matter how much individual or individual family fulfillment Fuller offered Americans with his highly mobile Dymaxion artifacts, including homes and cars, there was always an expected simultaneous fulfillment of global objectives. No matter how many young people adopted some of Fuller's ideas and inventions—as with Drop City—there was always a gap between those who dropped out of mainstream society and Fuller's credo of working within mainstream society. Nothing was to operate in a vacuum. Nowadays, insofar as experts are still relied on, it is a matter of consumers' demands and expectations, not their unquestioning trust.

True, America's consumer culture can be traced back at least to the late nineteenth century, as reflected in the rise of department stores throughout the country, but its pervasiveness and power have expanded along with high technology. The ramifications for utopianism today are extraordinary attention to material goods and to more individualized versions of "the good life" (no matter how many individuals ultimately choose the same version). The truism that one person's utopia may be another's "dystopia" surely applies to Fuller's visions now more than ever before.

Conclusion: The Real Value and Goal of All Serious Utopias

Fuller's utopias, like all serious utopias, are frequently misunderstood as supposedly scientific prophecies whose importance should be determined by the accuracy of their specific predictions. This view has grown increasingly popular in recent years, given the unprecedented electronic access to information and the consequent growth of supposedly objective forecasting. There is almost an inverse correlation between humankind's continued *in*ability to predict the future—save as shallow extrapolations from the present—and the growth of the forecasting industry. It was hardly an accident that in 1984 the head of the World Future Society, an otherwise highly respected scholar, chastised Orwell for his alleged failure to have predicted more accurately back in 1949 the "real world" of 1984—thereby completely missing Orwell's mission to try to prevent *1984* from becoming reality. Yet the principal value of utopias has always been and remains their illumination of alleged problems and solutions back in the "real world" from which they spring. Utopias should therefore be played back on the real world rather than be held up as crystal balls, much less viewed as scorecards.

Consequently, it is as important to determine the sources of Fuller's and his high-tech successors' popularity as it is to analyze their respective visions. This is why historical context is so critical. He and they are among the successors to earlier advocates of American "positive thinking" dating to Benjamin Franklin, and continuing in the mid-twentieth century with Dale Carnegie and Norman Vincent Peale. They are also the counterparts to such other contemporary positive thinkers as Stephen Covey, Robert Schuller, and the *Chicken Soup for the Soul* authors. The popularity of Fuller and his high-tech successors reveals much about Americans' and others' apparent obsession with the future. Despite professional forecasting's mediocre record, so many still feel compelled to try to uncover the future. That compulsion in turn reflects a deeper anxiety more than a superficial optimism, a profound uncertainty about the future that even our unprecedented access to information and to communication cannot overcome. The unceasing hype over our allegedly unique pace and extent of technological change has, I suspect, made many persons increasingly eager to figure out what comes next.

Equally misunderstood by many technocrats like Fuller is the critical role of politics in local, national, and world affairs and the inability to replace the messiness and unpredictability of politics with antipolitical technocratic thought and action. This is in part why one does not find in the writings of truly profound visionaries like Marx and Engels the kind of simpleminded technological determinism that colors not only Fuller's books but also those like Alvin Toffler's *Future Shock* (1970), Bill Gates's *The Road Ahead* (1995), and Nicholas Negroponte's *Being Digital* (1995). It is notable that Silicon Valley's leading figures and companies belatedly grasped this elementary fact of life. Only then did they set up offices in Washington, DC, and make regular visits there themselves. Compare this with the pathetic antipolitical performance of the Technocracy movement in the 1930s and its failure to gain major influence in American government.[15]

Revealingly, in the introduction to his *Inventions: The Patented Works of R. Buckminster Fuller* (1983), written shortly before he died, Fuller provided an illuminating end-of-life reflection entitled "Guinea Pig B." Here the normally bombastic, inscrutable, and self-centered utopian is surprisingly restrained, lucid, and modest. He characterizes himself more as God's humble servant than as God's anointed prophet: "I hope this book will prove to be an encouraging example of what the little, average human being can do if you have absolute faith in the eternal cosmic intelligence we call God." Even his familiar utopian rhetoric is tempered by the recognition that the world, alas, is still far from anyone's ideal: "We have reached a threshold moment where the individual human beings are in what I consider to be a 'final examination' as to whether they, individually, as a cosmic invention, are to graduate successfully into their mature cosmic functioning or, failing, are to be classified as 'imperfects' and 'discontinued items' on this planet or anywhere else in [the] Universe."[16]

Here, then, might be an appropriate starting place for reconsidering the life, times, and contributions of R. Buckminster Fuller.

Thinking and Building

The Formation of R. Buckminster Fuller's Key Concepts in "Lightful Houses"

Joachim Krausse

4

I

Design and discourse, although sometimes disconnected, form a unity in the work of R. Buckminster Fuller. They are complementary, like the tensile and compressive forces in his tensegrity constructions (fig. 4.1). His work would lose its coherence if one of the two aspects were eliminated. The artifacts he designed have always referenced and provoked this discourse, which Fuller developed continuously from the moment he began designing in 1928. His appearance at the Architectural League of New York on July 9, 1929, was symptomatic of this complementary relationship. At that meeting Fuller explained his Dymaxion philosophy to an audience of experts with a model of the Dymaxion house (fig. 4.2).[1] The free lecture not only introduced listeners to the design of his house but also helped its designer to complete and detail the provisional model by talking about the advantages that the house would provide for its imaginary users. In its unfinished state the model allowed Fuller to effectively develop his philosophy of the house, which went beyond the visible object. The model was only a temporary crystallization of a train of thought that the speaker tried to pull the audience into, successfully. Fuller's thinking, though, was not complete with mere verbal expression. It was rather supposed to be incorporated into the designs and to objectify itself in models, prototypes, and usable artifacts. He never left any doubt concerning this direction of thinking and the corresponding orientation of his discourse. During one of his most concentrated and visionary lectures, in front of the planning commission of Southern Illinois University in St. Louis on April 22, 1961, he outlined this position twice in short succession. He expressed his design philosophy as follows:

> During one-third of a century of experimental work, I have been operating on the philosophic premise that all thoughts and all experiences can be translated much farther than just into words and abstract thought patterns. I saw that they can be

Figure 4.1 Fuller with tensegrity models, Southern Illinois University, Carbondale, 1959. Source: Special Collections, Stanford University Libraries.

> translated into patterns which may be realized in various *physical* projections—by which we can alter the physical environment itself and thereby induce other men to subconsciously alter their ecological patterning.[2]

Distancing itself greatly from the common design theories of the time, this statement rejected the monopoly of verbal expression as a medium of thinking, as it was treated by philosophy and the humanities.

Verbal expression is for R. Buckminster Fuller neither the end in itself nor the final result. Rather, it is an intermediate stage or a different course that should be explored if one wants to reach the goal. The thinking is expressed in patterns, which can also articulate themselves differently than ordinary verbal forms do—more directly, as in the forms of articulation of the arts. Fuller, however, never intended to see himself as an artist but rather as a researcher who investigates the modes of thinking by different forms of articulation, first and foremost in the medium of geometry. This is why the subtitle of *Synergetics* is *Explorations in the Geometry of Thinking.*[3]

Other than in philosophical epistemologies, thinking is understood to be a continual elimination of the irrelevant—it is a process of refinement like in the field of metallurgy, not primarily a process of buildup like in constructivist theories. "Thinking," writes Fuller in 1963, "is a putting-aside, rather than a putting-in discipline. Thinking is FM—frequency modulation—for it results in the tuning out of irrelevancies (static) as a result of definitive resolution of the

Figure 4.2
R. Buckminster Fuller with his second Dymaxion house model. Meeting of the Architectural League, New York, July 9, 1929.

exclusively tuned-in or accepted feedback messages' patterns differentiability. And as the exploring navigator picks his channel between the look-out-detected rocks, the intellect picks its way between irrelevancies of feedback messages. Static and irrelevancies are the same."[4]

To make these patterns tangible and understandable, one has to create models. This is the task that Fuller has set for himself, and "Synergetics" is his way of achieving it.

II

An indispensable medium of articulation of this thinking about thinking is the graphic notation of these patterns. This is why Fuller took the advice of his friend and colleague Ed Applewhite to extend epistemology by the concept of *epistemography*.[5] Epistemography proposes a "geometry of thinking" and offers a tool for the criticism of persistently deceptive conventions of verbal terminology. Fuller illustrates this need with the help of an ineradicable building metaphor, the foundation, or that which is known as fundamental:

> There is also trouble with the word *fundamental*. It means foundational when there are no foundations . . . no two-dimensional planar base. The Earth and other objects are co-orbiting the Sun at 60,000 miles per hour and are gravitationally tethered to one another. The word foundation implies an impossible standing-still-somewhere in the Universe . . . on a solid and square or planar base. We may use the word *primitive* only to describe the initial self-starting conditions of awareness and think-about-ability of the minimum essential components of any evolutionary system's divergent or convergent considerability. Thus the primitive conceptual angle as one myopically viewed corner of the 12 corners of the minimum system has greater meaning than the expression *fundamental particle* employed by the high-frequency research physicists. The statements of this paragraph are strictly within the concerns of epistemography.[6]

(Fuller likewise criticized the word *sunrise* as a confusing term that suggested, quite inaccurately, that the sun orbited the earth.) The epistemography of *Synergetics* is an effective means of criticizing language and of analyzing terms but is in no way a simple discourse. *Synergetics*, after all, has almost fifteen hundred pages. This discourse is as necessary as it is complex since the reduction of the seemingly self-evident cannot be done with the key terms of common sense. "What humans have been experiencing and thinking of 'realistically' as dim 'somethings' or 'points' in a field of omnidirectional seeming nothingness now requires experimentally provable reconsideration, epistemography reconceptioning, and rewording."[7]

This determination to reconstruct language resulted necessarily in the fact that Fuller's most daring reflections remained cryptic for the unprepared reader. It has been noted more than once that this discourse, which was not always easily accessible in written texts and books, was very much understandable by an audience without previous knowledge if offered in the performance of a lecture. While transcribing the audiotapes for *Synergetics*, Ed Applewhite found out that this difference could not only be attributed to the discrepancy between verbal and written communication. It also has to do with the fact that the discourse without model (in the sense of epistemography) remains essentially incomplete. "One of the troubles with tape recording Fuller's talk," Applewhite writes in his book *Cosmic Fishing*, "is that so much of the message is conveyed in body English—hips twirling imaginary hula skirts of ball bearings, elbows pumping precessionally as pistons of an internal combustion engine. Often when he was dictating to me, or rather sharing an articulated stream of consciousness, he would become exasperated when I would not look at what he was doing with his hands."[8]

While trying to translate Fuller's work into a foreign language the interpreter of his writings becomes painfully aware of this deficiency, since the words are separated from the embodiment. To make sense of the written texts, one can only go back to the artifacts: drawings, models, and built structures, since the supporting body expressions are missing.[9] The difficulties of understanding can be overcome if one can assign the texts successfully to the corresponding artifacts. The discourse concludes only in these artifacts, and the artifacts, intended to be instructive, point beyond themselves toward the context of the comprehensive discourse. The artifacts altogether represent a more general idea concerning human relations to the world, insofar as Fuller's houses suggest a new relationship between the individual and the cosmic reality that the people do not yet perceive. Architecture mediates this relationship in a way that includes the sensorial range of perception and opens up the mind's receptivity beyond those limits that Fuller called the *house habits of thought*. His description of the world is unique because he is the only one in the modern era who investigates and clarifies the conditions by which we can develop a cosmic consciousness in order to modify our habitual patterns of living and thinking. This transcendental approach of making human beings at home in the universe tries to reconnect what has been so hopelessly disconnected by the highly specialized branches of knowledge. Fuller tries to trace the development back to the primary conditions of human living. In *shelter* he found the universal integer of regulatory principles. In that sense he named his comprehensive achievements a *universal architecture*.

III

William Kuhns remarked that "Richard Buckminster Fuller is a nineteenth-century inventor with twenty-first-century ideas. The fact that he lives in the twentieth century seems a dual anachronism."[10]

Fuller's way of working is indeed closer to the experiments of Alexander Graham Bell and Thomas Alva Edison than to the methods of scientific institutes. The goal (which he consistently pursued until his death) to deliver a design for a description of the world, a *Cosmography*,[11] as last published by Alexander von Humboldt with his main work "Kosmos" (1841) and by Edgar Allan Poe with "Eureka" (1848), also connected him more with the nineteenth century than with the twentieth. Fuller directed all of his research toward this *modelability*, which principally everybody should have access to. As such, his work was very much anachronistic since he worked against the conviction prevalent in the natural sciences between the world wars, that visible and tangible models had to be renounced in favor of mathematical formulas.[12] Fuller was successful with this seemingly obsolete thinking only very late, virtually after his death (1983). Its acknowledgment by the sciences came with the discovery of the C60 molecules, the third structural form of carbon after graphite and diamond, by Kroto, Smalley, and Curl (1985, Nobel Prize 1996). They named the cage molecule consisting of sixty atoms the *Buckminsterfullerene*, after R. Buckminster Fuller; the larger family of such molecules is called *Fullerenes*. This newly discovered structure of carbon initiated an unprecedented modeling boom in chemistry and molecular biology and gave nanotechnology a practical meaning.[13] The honor to be enlisted into the annals of science is unfortunately not attributed to a deeper reception of Fuller's *Synergetics* but to the simple fact of *pattern recognition*. While trying to identify the unknown structure of the molecule, Kroto and Smalley remembered Fuller's EXPO-dome, which both of them had seen at the 1967 world's fair in Montreal, and concluded that the desired structure must look similar to this dome (fig. 4.3).[14] Other, similar models were also considered, but it was necessary to confirm their guesses with existing models. Apparently nothing was as suggestive as the gigantic EXPO-dome, seventy-four meters in diameter, which Kroto vividly remembered having walked through. The dome, as a building, was furthermore a structural model that could be experienced from both outside and inside, a building where function and form were identical. The structure demonstrated the *patterning* that for Fuller creates the connection between thinking and building. In other words, a building can become a model for a new perception of space, a catalyst of research—an epistemological object of sorts. One of Fuller's unique achievements is that he designed all his artifacts as epistemological objects, that is, as occasions for the

Figure 4.3
Fuller and Sadao's EXPO-dome, Montreal, 1967.

observing and reflecting upon of one's own *house habits of thought*, as Fuller put it. He tried to free people of their structural prejudices. Therefore he advised his students to study chemistry:

> Particularly, I have urged them to learn what they can of Chemistry, for I feel that chemistry is basic structure, ergo architecture. . . . But what one learns in Chemistry is that Nature wrote all rules of structuring; man does not invent chemical structuring rules; he only discovers the rules. All the chemist can do is to find out what Nature permits and any structures that are thus developed or discovered are inherently natural.[15]

Who would have thought that architecture could one day give chemistry a helping hand?

IV

As we have seen already, architecture plays a major role in Fuller's attempt to achieve a cosmographical readjustment of habitual patterns. Architecture, for him, becomes a medium for redefining the relationship between humans and their environment, between thinking and doing, and between scientific-technical development and conditions of everyday life. In *Synergetics*, as a general, epistemographical systems theory, this is so much in the background that one hardly recognizes how this thinking relates to the problems of shelter and the philosophy of the house. In 1963 Fuller published four books and with this became known as an author and theoretician to a broad public for the first time. Some readers were astonished if not irritated at the wide scope of topics and the different forms of writing (essay, lecture, poem) that Fuller tried his hand at.[16] Had it not been for his international breakthrough in 1954 as the inventor and architect of the geodesic dome, and for Robert Marks's 1960 monograph that documented the experimental structures, as well as the realized dome constructions, the reader would not have been able to understand what Fuller was mainly engaged with.

Reyner Banham was the first architectural theoretician to put Fuller's early work into the context of modern design of the first third of the twentieth century.[17] After having read Fuller's books, Banham wrote in 1963: "But in all this we should remember that his idiosyncratic and effective mathematical facility is a direct product of his preoccupation with human shelter and is fed back into the shelter situation as a structural discipline and a methodology of environmental control."[18] This reminder of Fuller's central topic has to be taken seriously if one sees Fuller as an experimental geometer, cosmographer, or epistemologist because which model, but the house, still allowed a holistic approach to the sciences, technology, and economics within the frame of reference of personal experiences?

V

In his ample and continuous autobiographical discourse, Fuller always pointed out the turning point in his life in 1927. The autobiographical passages in lectures and books surely have the dual function of self-reassurance on the one hand and direct personal encouragement of his audience and readers on the other hand. These passages are not, however, free of mystification and legend. The literary scholar Hugh Kenner spoke of "Bucky's myths" not for the sake of denunciation but in order to promote an understanding of the poetic dimensions of such stories.[19] However mythological these stories might be, Fuller left a private archive that could hardly have been more complete, and these papers may be used to check his honesty. But there are of course deferrals, changes, and simple forgetfulness. Forgotten, and obviously never looked up again, was one of his very first attempts to achieve clarity over the design of an innovative house, found in a seventy-four-page typescript, together with drafts and different bundles of small slips of paper with notes and sketches and one bigger pencil drawing with the preliminary sketch. Fuller calls the project, at its time of first conception in January to March of 1928, *Lightful* or *Lightful Houses* (fig. 4.4).[20]

Figure 4.4 (opposite) Sketch for "Lightful Houses" (early 1928). © Estate of R. Buckminster Fuller. All rights reserved. Used by permission.

The documents pertaining to this work show the true beginnings of design in the sense of being Fuller's first attempt to articulate his housing project.

Time, Light, Transportation

Life, Lightness, Cheer, Strength, Color

Loveliness, Cleanliness,

Godliness, Truth

LIGHTFUL HOUSES

When one compares the text of "Lightful Houses" with the *4D* book, it becomes clear that the former provided the basic text and rough draft for the latter. The *4D* book was developed by cutting the original typescripts and by adding new text. The work on these small conceptual sketches seems to overlap with the hurried preparations for a patent specification in February to April of 1928. They all document the search for a valid structure for the house on a mast. One reason for his forgetting the entire complex could be the fact that at this early stage the ground plan was still square, which from a later perspective turned out to be a major mistake. If one puts this into the context of how carefully Fuller investigates the geometry of a hexagonal ground plan in his *4D* sketches and the fact that he thereby discovers the possibility of symmetrical, radial growth (from the center to the periphery as a time dimension), it becomes clear that the seed of *Synergetics* is found in the structure of the Lightful, 4D, and Dymaxion house.[21] During the design process, Fuller realizes that his ideas of organic growth, of the dimension of time and an unfolding life, are at odds with the square and rectangular grid, with the cubic room and the right-angular module, which seem unavoidable for the traditional building production. This will remain a constant drive for the search for how nature is able to let the processes of growth take such a manifold of shapes. It becomes a central question of *Synergetics*: how does nature coordinate?

We should consider the large pencil drawing, which contains the writing "LIGHTFUL" in the second concentric circle, as programmatically as well as characteristically for its author (fig. 4.5).[22] It shows how Fuller approaches the problem of the house that is to be designed. It does not show the house but the world—the world for which the house is meant. This planetary-extraterrestrial perspective, in which the earth globe is seen and put right at the center, can be found again and again in his later work—the *Dymaxion World Map*, the *Spaceship Earth*, and the *World Game*.

One cannot look at this drawing without being touched. This might have to do with the fact that Fuller has drawn four highly visible symbolic objects on the corners of the paper, which form the frame of reference under which everything is placed. They mark, so to say, the personal coordinates.

The four objects are, in the order of reading: heart, sun, church, and baby. They all have a suggested halo. The lines of writing that connect the symbols form a frame:

- THE EXQUISITE LIGHT / SLOW MATTER
- TIME FELLOWSHIP PRODUCTION
- TIME METAL MECHANICS

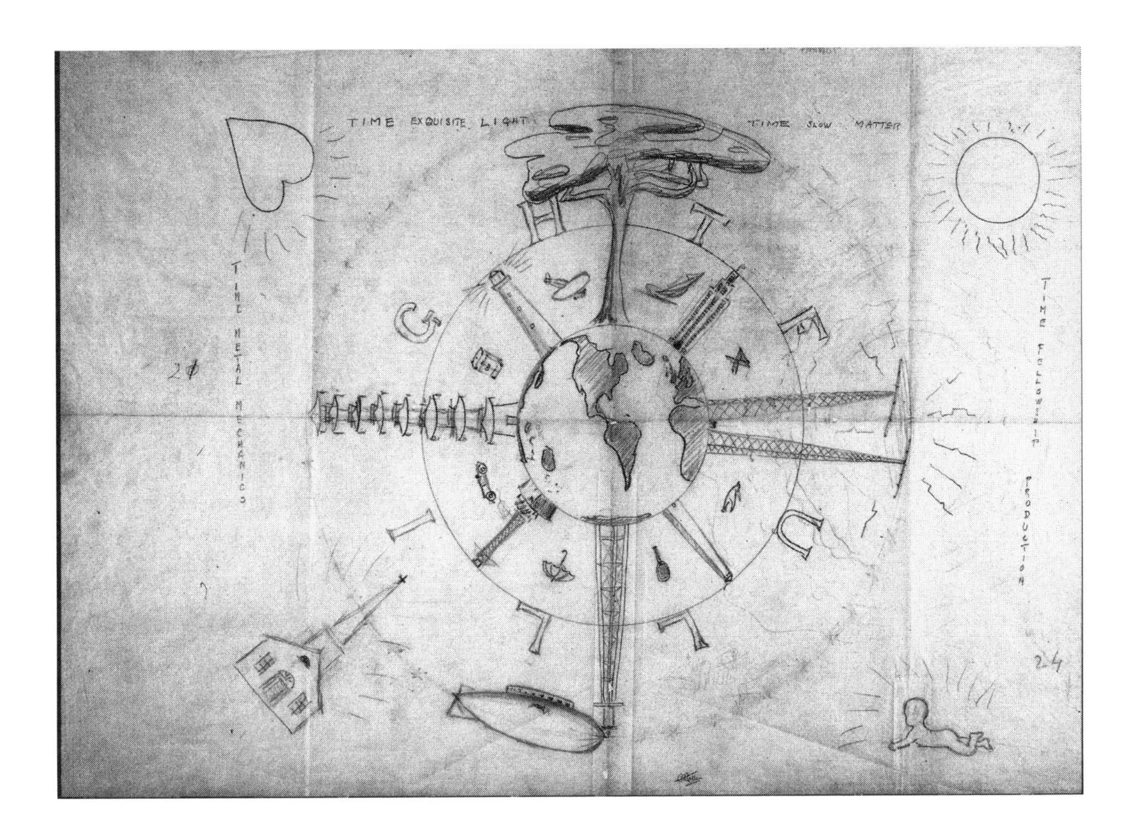

Figure 4.5
Lightful,
Fuller's earliest programmatic drawing (early 1928).

The baby (bottom right) gives this frame of reference a personal note with a strong autobiographical relation; Fuller's daughter Allegra is six months old at this time. The father is unemployed, at home (in a cheap apartment in Chicago), and has time to watch the baby grow up. It is being weighed and measured, and Fuller sets up a data collection in a folder that is titled "Baby's Record."[23] One can assume that the baby has become a central, emblematic figure of thought in the process of design since it reappears on numerous drawings and is called *President of 4D* in private documents (fig. 4.6). The later discourses on the shelter's primary function to protect *the new life* are derived from this context of experience.

A comparatively small Earth is found in the center of the four emblematic corner points (love, light, new life, and religious spirituality). Vertical structures that spread like spokes or rays in all directions—*omnidirectional*—originate from the surface of that globe. The larger ones, forming two concentric circles around the main horizontal and vertical axes of the globe, show a dominating tree and a downward-pointing, tetrahedron-shaped mooring mast with an anchored airship at the top. This mooring mast—a construction

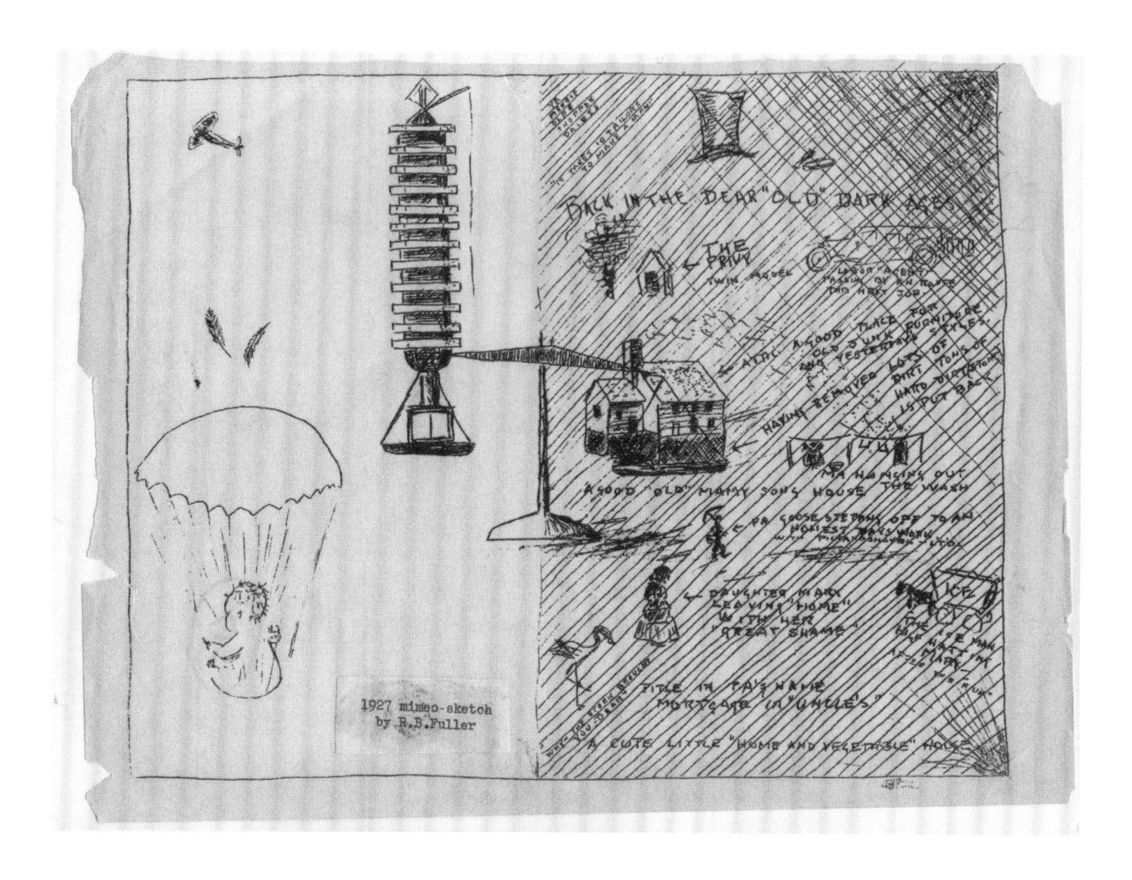

Figure 4.6
The baby as *President of 4D*, mimeograph drawing (1928, predated by a later added inscription).

of the British engineer Barnes Wallis—becomes the example for the construction of the house's support and the starting point of Fuller's reflection of the structural superiority of the tetrahedron. The tree, because of its dominant size, at first recalls the archaic myth of the tree of the world or the tree of life. He then transforms it into a structural-functional model for patterns of supply, as well as for patterns of the distribution of tension, which can be technically adapted. This motif reappears in his later work. The Dymaxion house, for example, is also called "a house like a tree." Fuller finds his examples or conceptual models for the design not only in the artifacts of civilization but also in nature. The horizontal left side of the drawing (geographically pointing toward Asia) shows a pagoda with its hexagonal floors, surely also an inspiration for the 4D Tower. On the right side one sees diverging high-voltage power masts, which introduce the topic of the world's energy supply. In the inner concentric circle one finds the American skyscraper, which Fuller thoroughly analyzes in "Lightful Houses" in terms of its weight-to-volume ratio. One also finds an obelisk (as a gnomonic meter of time) and a network tower of the

U.S. Navy, a construction that goes back to the ingenious Russian constructor Wladimir Šuchov and that shows the resolution of matter in a network of interlaced rods. There is also a lighthouse, a signal-sending device for navigation and orientation. The catalog of these eight vertical structures shows that Fuller's focus regarding building construction evolves during the investigation of the supporting function. Pillar, pole, and masts embody the evolutionary connection between buildings and humans. Just as anthropoids have evolved to walk upright and stand erect, and as infants learn to stand and walk alone by leaving their mother's helping hand, so, too, has the erection of buildings evolved with the help of supports. Fuller underlined the importance of this opposition to gravity for life and for the cultural history in his essay "Vertical Is to Live, Horizontal Is to Die."[24] It is distinctive for Fuller that the classic interpretation of the supporting function in the model of the *column* does not appear in his catalog of "vertical features." We should keep in mind that the canon of proportions has been determined in the orders of columns from antiquity until the nineteenth century. This canon used to be the central piece of academic architecture education. Its omission must be interpreted as an act of provocation. The catalog of vertical structures is at the same time an example of "putting aside of the irrelevant." The topic of the mast reappears in Fuller's architectural work as one of resolution and temporalization. In a rapid sequence of pictures of his houses we could see how from the Lightful house to the *Wichita Dwelling Machine*, everything is organized around a mast as central axis. Geodesic domes as clear-span constructions do not need supports anymore. The mast only appears temporarily, as a lever or crane during the erection. For the rest, the mast is dissolved in the network of rods of the geodesic web. Fuller was always suspicious of the column and the corresponding structure of the beams, owing to the illusion of permanence connected with their classic tradition. Instead, Fuller focused on the cyclical meandering of forces in *regenerative circuits*. More strength could be derived from feedback loops than from substantial permanence.

Returning to the drawing, the space in between the vertical structures shows objects of everyday life, which represent paradigms, certain principles of design. The tennis racket, for example, demonstrates the strength of a stretched net in a frame at very low weight. The everyday object becomes a cause for thinking about the represented principle; it is a model that can be adapted and made useful for the design. Fuller's idea to replace massive ceilings with decks, which are spanned like nets, originates from the tennis racket.

Although this series could well be continued, I have used this microanalysis to show how Fuller's conception and analysis, during the process of his very first design phase of this house on a mast, is organized in such a way that the partial solutions can be fed back again and again with the systematically structured whole. Building a house and the description of the world based on this level of experience are becoming the interdependent impulses for a human strategy of bettering the world.

VI

The *Lightful* drawing does not stand for itself. It rather reveals its meaning only in the context of the discourse that Fuller unfolds in his text "Lightful Houses," as well as in the sketches, text drafts, and notes relating to this complex.

His focus on internally supported vertical structures in the drawing, for example, demonstrates his approach to the problem of the house by designing from the inside out, instead of vice versa. In "Lightful Houses" he states: "One cannot design from the outside in. There can be no character unless we design from the inside out. The surface must express the interior functionalism and life."[25]

This maxim explains the determination with which Fuller initially turns to the load-bearing structure and the supply functions of the house. We are reminded at the same time of how the topic of the "inside-outing, outside-inning" is becoming a leitmotif of his geometrical research. His impressive discoveries such as the Jitterbug-transformation or the transformation of geodesic grids in *Noah's Ark #2* are marked by such inverse transformations, which, in mathematics, remained unrecognized for a long time.[26]

Fuller's principle of tensegrity is also based on the thought of inverting the arrangement of the tension and compression components of a structure. One can say without exaggeration that Fuller's approach to the problem of a space-time order originates in this thought of inverting the inversion or the "inside-outing" as he put it.

Designing from the inside out has numerous aspects and contexts in "Lightful Houses." The most important one is probably the analysis of life and of living itself, because this is a constant source of inspiration for all of his designing. The dealing with growth, unfolding, transformation, and reproduction leads from the living creature and nature to the cultural techniques and artifacts. To label Fuller a technocrat, as has been done often, only because his solutions want to exhaust the technologies of the time without compromise, misjudges the role that nature and the organic world have played in his thinking (fig. 4.7).

I have already pointed out how much the circumstance of the birth of his own child in July 1927 contributed to the change in R. Buckminster Fuller's

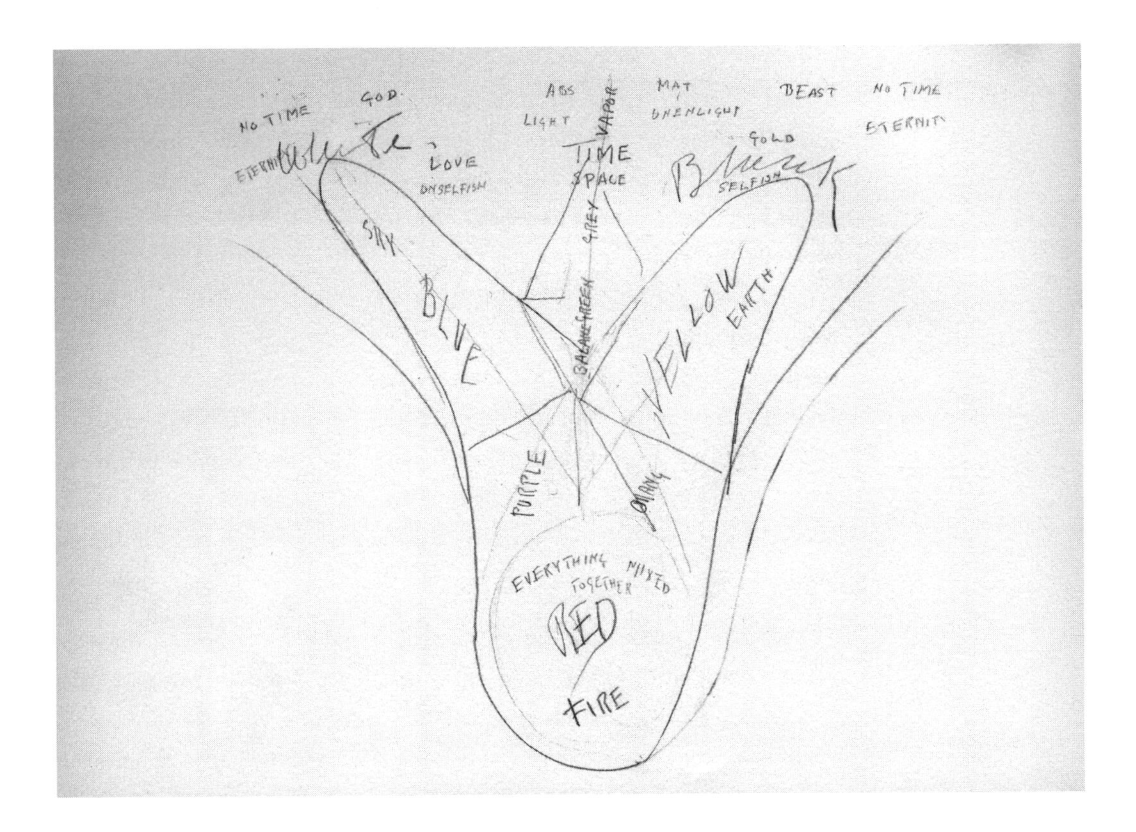

Figure 4.7
Ideographic models of growth: untitled sketch from the Lightful papers (undated [early 1928]).

life in the winter of 1927–28. This autobiographical aspect should not be underestimated, because it continuously appears in the notes of that time, and it interweaves the personal-familiar with the objective-general. One example is the diary that Anne Fuller, expecting the baby, started on August 7, 1927, and that she and her husband kept until the end of March 1928. One entry by Anne Fuller reads: "RBF terribly inspired by life. It is source, design & motivating power. Basis of planning & thinking etc." This entry from February 2, 1928, falls right into the period of feverish work on the project "Fuller houses," as it has been called internally. The statement, as well as the entire diary, shows how impulses from the private circumstances in life connect with other stimuli and are being implemented into the philosophy of the house.[27]

The reflection of life and growth in general can also be found in the idea sketches that belong to the Lightful-complex. Some can be understood as abstract ideograms, others as idea sketches for the house's design. Among the ideographic sketches one finds a germlike or calyxlike depiction, which shows the unfolding of sprouts or leaves as a fanning out of the spectrum of colors around a central axis of time.[28]

Another one shows extending circles of waves around a center with the title "The Abstract / Truth." From the center, a cone-shaped bundle of five rays, naming the five senses, is directed toward the circles of waves. These rays are mediating between the truth and the waves. The metaphors and images of thought used in the central passages of the text "Lightful Houses" very much correspond to this. Fuller interprets them abstractly but also as concrete models of elementary patterns of motion from which he then gains a space-time construction ideal.

I would like to talk about two of these images of thought in more detail—namely the "fountain of life" and the radiating "expanding sphere"—because they are very important to Fuller's later work. They also show which traditions his thinking is following. "Lightful Houses" culminates in the passage where he draws his conclusions out of the maxim of designing from the inside out:

> We will have arrived at our new artistic era of architectural expression, when our buildings will have lost their last trace of feudalistic depression: when we arise in our buildings in concentrated area of compression in opposition to gravity by means of mast or caisson reach out in space from the vertical by tension and compression, compression diminishing as we fall off from the vertical, until we finally flow downward in direct tension. Then will our exteriors, hanging from the outward flow of the top like a great fountain be full of lithsomnes [*sic*], light and color.[29]

This astounding statement, by emanating from an aesthetics and ethics of architectural expression, seems to fit almost seamlessly into the contemporary discourse on architecture in America, masterminded by Louis Sullivan, to whom Fuller refers directly in the following sentences. The passage in its expression and verbalization does however show Fuller's own approach to building and to the house. What is initially striking is the fact that the introduced figure is imagined out of a process of movement, a process that can be understood and performed with the human body using gestures in a choreographed way. This embodiment helps everybody to experience the figure. But this is not all: Fuller's poetic description transports us right inside the flow of movement of ascending, spreading out, and descending—as if we are a part of what executes the movement or of what happens to a moved particle. This is how we feel—almost physically—the tensions that emerge from this movement and how the forces in the flow are inverted from tension to compression. The author finds an aesthetic solution for the imagination of the stress ratios and their transformation, which a static observer cannot immediately understand and which are, we might add, foreign to static thinking.

Fuller's description of the fountain is poetic, first of all in the understanding that he adopts from Ralph Waldo Emerson: "Poetry means saying the most important things in the simplest way."[30] The idea of the house that is to be designed has nowhere been expressed in such a concentrated way as in the author highlighted passage in "Lightful Houses." The fountain passage unites the ethical, aesthetic, functional, and structural conceptions of the author in one image, an image that has to be figured out. Its ambiguity demands that it be explained in concrete terms, but it already offers an indication of how a structurally functional model can be developed out of a metaphor.[31] The verbal concretion therefore reads in "Lightful Houses": "The basic idea of the construction is that all elements shall be suspended from above rather than rest upon supports from below."[32] The dynamic image of the fountain gives Fuller the freedom to detach himself from the professional categorization of architecture, the division in facade, body, load-bearing structure, and so forth, or from the building elements of wall, roof, and ceiling. This is how he arrives at a new designation of the house, as an umbrella and membrane. Here we see an essential precondition for the philosophy of *environment controlling*, which Fuller first develops in his 1938 book *Nine Chains to the Moon.*[33] The implications of this concept on the spacious roofing and climate-controlling skin that Norman Foster, Nicholas Grimshaw, Jörg Schlaich, Renzo Piano, Shigeru Ban, and many others are building today have so far only rudimentarily entered architecture theory and history.[34]

When, for example, Mies van der Rohe's architecture has been correctly characterized as *skin-and-bone-architecture* because he has drawn the radical consequences out of the skeleton building system and the possibilities of the curtain wall facade, then one could describe Fuller's building constructions to that effect that they are trying to get rid even of the skeleton itself in order to become mere membranes. The resolution of the skeleton can be described in two ways: first, by making the mast or supports temporary in the form of the building crane, and, second, by transferring the supporting function into the network of rods of the cellular trusses that subdivide the skin. To create the best ratio between the system (the house) and its environment, the membrane, by opening and closing the cells or facets, can be used to regulate the light, temperature, and humidity to optimize the heating, the lighting, and the airing. The most impressive realizations of *Geodesic Domes: The "Climatron"* in the botanical garden in St. Louis, Missouri, of 1960 and the EXPO-dome at the 1967 World Exhibition in Montreal are based on the concept of intelligent *environment controlling.* There is a complete separation of the building's skin and the interior structure, including the floor areas, the hallways, the ramps, the escalators, and so on—all this is now *becoming interior architecture.* One can say that the intelligent skin of the dome represents the *interface* between the shelter and its environment. It equally abolishes facade, roof, and outer walls and substitutes them completely.

The fountain figure even acts as an accoucheur for this concept of the completely free inner horizontal as well as vertical room organization, as it manifests itself in the EXPO-dome. It is not sufficient to characterize the geodesic domes as lightweight clear-span structures only. This is why Fuller has turned his attention to the inner and outer flows—of air, water, energy, material of supply, and disposal, as well as to the human procedure of action. Hence we find the fountain image from "Lightful Houses" again almost two decades later, in the description of the "Wichita House," this time relating to the functions of the house technology: "With the central vantage point for generating air, light, sound and work services, we discover that those services when in operation describe fountain-like flows upward, outward, downward in all directions with concentric flow for recycling below. We discover also that this fountain flow can be reversed but in either case, maximum coverage with least distance is effected."[35] In the Wichita House Fuller examines especially carefully the mostly natural air circulation (fig. 4.8). The air circulation is a result of the almost hemispherical shape of the house, the inlets and the aerodynamic extraction

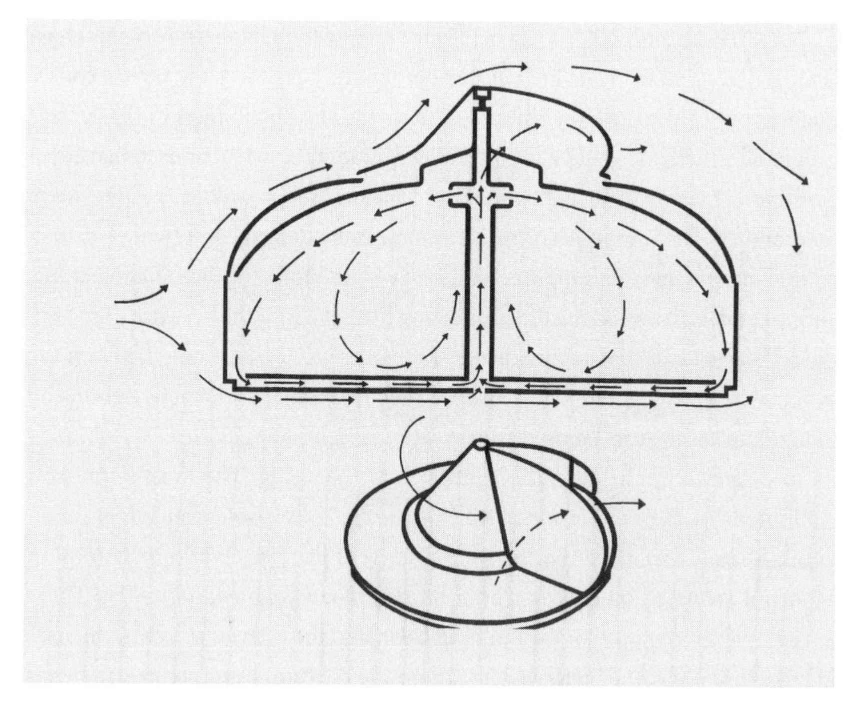

Figure 4.8
Dymaxion dwelling machine, ventilation of the Wichita House (1946).

fan. The circulation shows a fountain figure just like it would be in the natural atmosphere. The ingenious heating and airing concepts of the geodesic domes are pursuing the discoveries made by the Wichita House.

In 1951, after having finished his work on the geodesic grids, Fuller develops, together with his students at North Carolina State College, a new type of building for a cotton mill (fig. 4.9). The name *Fountain Factory* is accurate because the entire treatment process of the raw material is organized in this design. The raw material is transported upstairs inside a central supply shaft; it is then distributed outside and falls through the eight stories from one treatment stage to another in order to land as the finished product on the bottom floor, from where it is then carted away.[36] It is decisive, that for the very first time, the enclosure of the geodesic dome is completely separated from the inner superstructure. The ceilings of different radial lengths are light, planar, and open-frame trusses of the octet truss type, one of Fuller's three key inventions. They are anchored on a hexagonal shaft and suspended at the periphery, from the top of the dome. The three-quarter sphere carries only itself, has a climate-controlling skin, and offers all possibilities for a free inner organization. Its spherical shape is climate active in the sense that it allows and maintains the reversible airflow in the flow pattern of a fountain and therefore initiates an

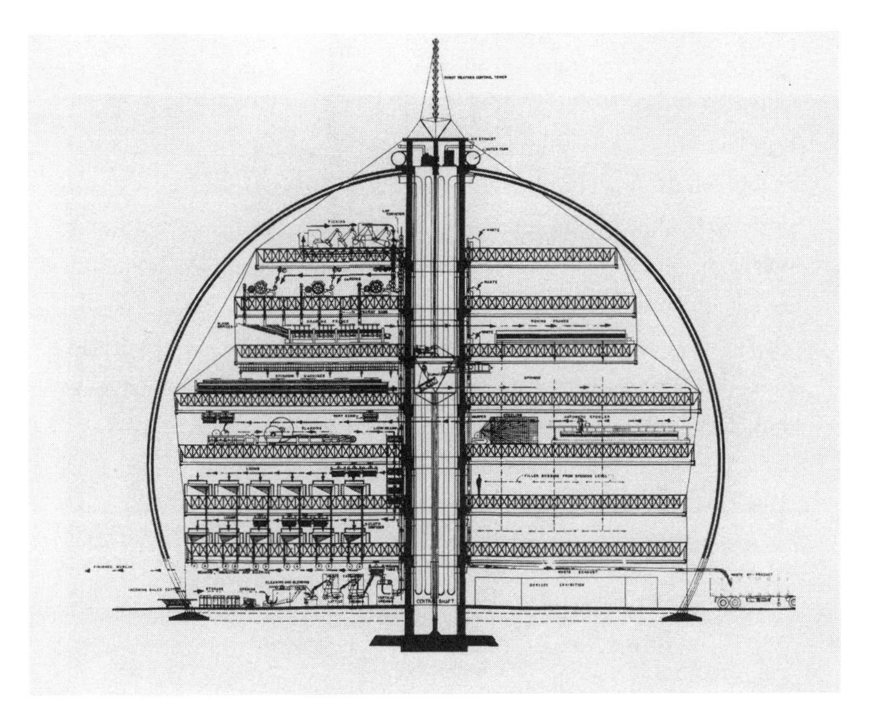

Figure 4.9
Fountain Factory: the 90 percent automatic cotton mill, project at North Carolina State College (1957).
 Source: Special Collections, Stanford University Libraries.

independent air conditioning. An eminent ecological argument has now enriched the structural criticism of the rectangle and the cube in architecture: "The fountain flow is appropriate for maintaining relatively warm atmosphere flow in winter, and reverse fountain is most efficient in maintaining relatively cool atmospheric flow in summer. In neither of these fountain flow cases does energy set up a chaotic echo system as we find it doing in the indiscriminate, cubical, squash racquet court shaped chambers in which we now live."[37]

We can see how often the image of the fountain reappears in Fuller's research and designs over a long period of time. Again and again, it is given new meanings as a model for a number of subfunctions that have to be coordinated organically—as is done by a living organism.

Fuller's unique way as a thinker and designer is not characterized by "form giving," but rather by examining how dynamic systems interact with their environment in order to favorably regulate them—or better—to let them favorably regulate themselves. This is why he devoted great attention to the studies of natural systems and described them in such a fascinating manner. Out of these analyses and descriptions, he developed a notion of the house as a system of *environmental controls*, and he did so independently of the natural sciences. Here is an excerpt from the lectures that Fuller gave to his staff of the Wichita House and for the technicians of the Beech Aircraft airplane factory, in Wichita, Kansas, where the prototype house was built:

> As a fountain of water is seen to operate freely in space as a system, or as light outdoors in the night creates a hemispherical system of illuminated space by atmospheric refraction of light, so also do these other dynamic functions of heat, light, air, sound and smell constitute natural systems of physical phenomena so that our hemispherical house is seen to afford only an isolating enclosure which complements the flow and systematic refraction angles and protects them from disturbance by dynamic conditions exterior to the house—as does a camp chimney protect the flame or an electronic tube protect the free functioning of its central element. The principle demonstrated by the boomerang refractions in all directions articulated by our coincident energy systems of light, heat, air, sound, smell, etc. positioned at the center of our house. In this way our house is dynamically faired (if not more so) as is an airplane, in order to induce a little parasite drag internally and externally to all the slip streams of dynamics as can be measurably arranged. Thus a minimum of energy provides a maximum of controlled service performance.[38]

When R. Buckminster Fuller met the biologist Ludwig von Bertalanffy in the early 1960s, he noticed a complete correspondence of his theory of natural systems that he had elaborated by modeling his house over decades with

Bertalanffy's *general systems theory*. The general systems theory had developed out of the necessity to explain the organism as a holistic system in interaction with its specific environment.[39]

Fuller's house-thinking, continuously developed between 1928 and 1967, which was by no means completed with the EXPO-dome, suddenly found itself on the point of intersection of the new "super disciplines" of cybernetics, ecology, and general systems theory. The house was like no other artifact, a holistic model that had made the "organic" integration of subfunctions and subsystems the task of the design. R. Buckminster Fuller experimentally researched and demonstrated this with a completely independent approach, arriving at a practical understanding of natural systems. For this, he will have to be considered one of the great pioneers of systems theory by the history of science.

VII

The close connections between thinking and building, discourse and design in the unfolding of R. Buckminster Fuller's lifework can best be understood in his central images of thought from "Lightful Houses." Instead of the fountain figure, the similarly prominent image of the radiating "expanding sphere" could have been used. I have investigated this elsewhere.[40]

Fuller's texts seemingly offer no direct hints as to where the images of thought, which became prolific so early, were coming from. The opposite, however, is the case in his famous autobiographical essay "Influences on My Work," which was at first sent as a letter to the young and unknown British artist John McHale on January 7, 1955. Fuller talks about his experiences as a child and adolescent, as well as his training and service in the U.S. Navy, with no mention of intellectual influences by literature or philosophical texts. He does, indirectly, however, give highly visible clues of the spiritual tradition that he is following. It is American transcendentalism, with whose literary-philosophical heritage he familiarizes himself at this very turning point in 1927–28. In the middle of his work on the great design for the future, he discovers the double bond with the circle of American thinkers and poets that formed around Ralph Waldo Emerson in Concord, Massachusetts, around the middle of the nineteenth century. Fuller studies Emerson's essays and discovers that his own great aunt, Margaret Fuller-Ossoli, played a central role in this circle and especially as a partner in dialogue with Emerson.[41]

By including a text of Margaret Fuller's in his book *Ideas and Integrities*, Fuller proves the importance of the spiritual heritage of his aunt Margaret. The short text forms the third chapter of the book, where a very revealing passage is printed under the title "Margaret Fuller's Prophecy." The text links the beginning of industrialization in America to the development of democracy and of a genuine American culture. Margaret Fuller critically asserts that—in 1842—it has not yet come to an independent American literature but that the European literature is only being imitated. The day of its independence would only come if

> the fusion of races among us is more complete. It will not rise till this nation shall attain sufficient moral and intellectual dignity to prize moral and intellectual no less highly than political freedom, nor till the physical resources of the country being explored, all its regions studded with towns, broken by the plow, netted together by railways and telegraph lines, and talent shall be seen till from the leisurely and yearning soul of that riper time national ideas shall take birth, ideas craving to be clothed in a thousand fresh and original forms.[42]

These lines from the past must have very much touched R. Buckminster Fuller as he read them, since he was in the process of finding his own position and was defending himself against cultural hegemony from Europe—this time not in the field of literature but in the field of building. The year 1927 marks an international breakthrough of the modern architecture that was successful in many European cities. The American architecture magazines in 1927 and 1928 extensively reported this.

The English translation of Le Corbusier's *Vers une architecture* (1923) was also published in 1927.[43] The importance of this book for Fuller is evident in his private papers and in his list of references for the 4D book. Here one finds numerous entries of technical literature but also writings by Henry Ford, Bertrand Russell, and Francis Bacon, as well as essays by Frank Lloyd Wright, Ralph T. Walker, and Ralph Waldo Emerson.[44] The list of references provides information about Fuller's readings until May 1928. However, it seems that he lacked the time to finish it before he distributed the *4D* book in May 1928 since the list was not included in the publication. But here, as well as in the *4D* correspondence, we see that Fuller was an eager reader who was well informed about contemporary architecture and technical developments, and who was determined to pursue his philosophical interests. In other words, Fuller was open to outside impulses and suggestions as he worked on his house project between December 1927 and May 1928. This, however, corresponds little with the image he created in "Influences on My Work."

It is very likely that he took great care in assuring himself of his own spiritual traditions, in the flood of information about revolutionary innovations of all kinds, especially architecture. And here, in the readings of Emerson and in the discovery of Margaret Fuller's work, he again finds the "images of thought" in the transcendental metaphors, which he can interpret and execute as models of natural systems and constructions. At the same time, these metaphors warn him to beware of imitation and of following trends and fads. For R. Buckminster Fuller, the decisive and lifelong valid sentences can be found in the quoted "Prophecy" by Margaret Fuller:

> The truth is the nursing mother of genius. No man can be absolutely true to himself, eschewing cant, compromise, servile imitation, and complaisance, without becoming original, for there is in every creature a fountain of life which, if not choked back by stones and other dead rubbish, will create a fresh atmosphere, and bring to life fresh beauty. And it is the same with the nation as with the individual man.[45]

Fuller felt obligated to this heritage, but he did not preserve it via the literary-philosophical avenues that the earlier transcendentalists preferred. He rather translated this thinking into his tangible work, work that points beyond the contemporary ties and seems like a prophecy itself—not only an American one but a global, planetary one.

5

"Spirit House" and "Steppenwolf" Avant-Garde

American Origins in the Dymaxion House Concept

Claude Lichtenstein

> In architecture "form" is a noun; in industry "form" is a verb. Industry is concerned with *doing*, whereas architecture has been engrossed with making replicas of end results of what people have industrially demonstrated in the past.
>
> Fuller, *Nine Chains to the Moon*

Seventy-five years after its conception, R. Buckminster Fuller's 4D-house (Dymaxion house) venture retains its unique character.[1] Never since has architecture been attacked so fundamentally, and reinvented so genuinely, as it was with Fuller's intellectual thunderbolt. No wonder his project has remained unrealized, not only as a specific project but also as a general strategy for housing. Few events in history of architecture can compare to the 4D-house project of 1928. It remains an alien element in architecture, an "odd case," as was its author, R. Buckminster Fuller, among architects, whose very profession the man himself put into question in the most radical way.

The 4D house, however, raises some questions. What is the character of this event? Is it a specific project, or is it an imaginary model? Or is it a highly suggestive appeal to achieve a more flexible understanding of the nature of dwelling? Fuller might have laughed about these questions. In 1928 he was convinced of the imminence and importance of his proposal. But after 1930 he might have learned to deal with it in a more general and didactic way.[2] Fuller's attitudes were dynamic rather than definitive. He started the 4D-house project to discover a generally valid formula for a wholly alternative standard of building activity. His project dealt with a horizon different from the usual one, and as it is the case with a horizon, it recedes to the same degree as one proceeds. While his goals remained the same for more than half a century, he continuously revised the repertoire of his tools and his ideas about how to achieve the envisaged goals.

Like Henry Ford, who actually made the profession of custom-built coach building obsolete, Fuller advocated a completely new understanding and practice of house building and dwelling that would harness the efficiencies of the machine age. Yet, although Fuller advocated technical means and mass production, his 4D house also had deep personal and spiritual inspirations.

As such, the Dymaxion house inevitably has a double character. It aims to be impersonal and objective, yet, because of its outstanding character, it witnesses an extremely personal (however plausible and consistent) view. Having remained a project—Fuller's first and thus maybe his most important project and his launching platform as an author—it is the thrilling document of an individual trying to set new standards. This is a typical phenomenon: the quest for universal qualities can actually be a highly personal initiative. Faults do weigh more heavily if they are due to the universalistic approach, but they fade out if we see the personal signature in a proposal. Le Corbusier, for instance, is considered by some to be an inexhaustible source of inspiration for exciting ideas. To others, who are attached rather to measurable effects than to impulses, he is responsible for "having murdered the city." And R. Buckminster Fuller, was he the orbiting apostle of unleashed technological progress and indestructible optimism? The answer requires some qualification. Yes, he believed in progress, and he intended to "make the world work" by harnessing technological and scientific progress; but he was aware that this alone could not replace human progress. Technological progress should be used only as an aid to human progress. The Dymaxion house was his motif to crystallize a new model of existence.

Fuller's Radical Questioning of Architecture

Strangely enough, design is widely approved as creativity applied to mass production, and it profits from a wide acceptance of this dimension, whereas architecture is often considered the art of the custom-built single item. The fragile acceptance of handicraft is most warmly appreciated in the domains of jewelry, of haute couture, and of architecture. In addition, the world of architectural publicity and discourse is tightly (however unconsciously) attached to the idea of uniqueness, with the architect as the passionate hero. Each building, each home claims to be a single-case invention. From TV ads to film sets and the policies of the architectural press, the glamorous custom home is the symbol of a successful life in financial abundance. But, alas, this is not the '*Great Combination Revealed Awaiting the Click at Each Turn*' that Fuller announced in his mimeographed *4D Time Lock* papers in 1928.[3]

Fuller always addressed himself toward humankind, and his job was to promote inclusive standards rather than exclusive extravagances. More radically than anyone else, he proposed the turnover of architecture from a handicraft issue to a true industry. But did he, in 1928, want to reshape the discipline of architecture? I do not think so—he was not a discipline-bound thinker. His intention was not to reshape the grammar of architecture; rather, he proposed a revolutionary antithesis of building. Unlike Le Corbusier or Gropius, he did not propose the industrialization of the building process according to an established repertoire of architectural themes, but he changed the idea of "architecture" according to his ideas about industrialization in house building. His ideas were imported from the automobile industry, from shipbuilding, and from aircraft technology. This venture led him gradually from the form-based right angle (the seemingly stable, in fact unstable, right angle) and later (from 1950 on) away from the tyranny of verticality to an omnidirectional structuring. His was the shift, generally spoken, from a shape-oriented to a structure-oriented understanding of design. The first step was the 4D house, with its revolutionary replacement of masonry and brickwork by metal and plastics.

Only a non-architect could work out a scheme as Fuller did with the Dymaxion house. Fuller himself identified this fact later (1946) in a talk given to technicians, remarking, "As engineers you certainly understand that man is born inside the frame of scientific measurement reference, therefore it is impossible unless he gets 'outside' of the whole phenomenon house for him to be very critical of his performance standards."[4]

R. Buckminster Fuller himself had gotten "outside" and had rejected the common assumptions about architecture. He had become familiar with the ruling mentality in the building business when he was the director of the Stockade Building System company from 1922 until 1927. He had witnessed, surveyed, and aided in the construction of 240 houses (along the East Coast and in the Midwest) according to the building method developed by his father-in-law, James Monroe Hewlett, and himself. An album with "Stockade" photographs shows that each house looked different from the others. This was the point that to Fuller became the pivotal problem, the question of a flexible system to facilitate the handmade process of building or the "teleological" approach. He mentioned his "initial teleologic preoccupations" and named his work "their resultant proclivities."[5] Teleological to him was "a within-self communicating system that distills equitable principles . . . from our plurality of matching experiences."[6] This led him to the design of a mass-producible house in a production mode analogous to that used in the automobile industry.

Fuller was not alone in wanting to replace, after five thousand years of building history, heavy walls and ponderous beams with a heavy-duty structure made of modern materials. This was the core idea of "modern architecture" in Europe and Russia after 1920.[7] The Swiss avant-garde "ABC" group coined contemporarily a visual analogy formula much like Fuller's polemics. *ABC* was a periodical published in Basel between 1924 and 1928; the authors were Emil Roth, Hans Schmidt, and Mart Stam. Russian constructivist El Lissitzky, who lived in Switzerland in 1924, hoping to cure his tuberculosis, was also associated with the group. They opposed the traditional and the new building practice with the amazing illustrations of "building times kilograms equals 'Monumentalism'; Building divided by kilograms equals 'Technique.'"[8] The thought is close to Fuller's "teleology" keyword and to his "doing the most with the least."[9] But there remains a difference between how Fuller and Europe's avant-garde questioned architecture, although the most radical architects in 1920s Europe had arguments similar to Fuller's. For instance, Hans Schmidt and ABC proposed that the building activity be focused at the maximum of impersonal character and of artistic "indifference," and he stressed, in his action-based understanding, the difference between (static) "architecture" and (dynamic) "building" much like Fuller did;[10] however, he remained within the reference frame of—in Fuller's words—"cubical" architecture.

The Dymaxion House as an Outpost

Fuller, who toured in the Unites States in 1929 with his crudely cobbled-together model of the 4D/Dymaxion house, was undoubtedly an outlaw, and the successful (sometimes European-trained) U.S. East Coast architects may have been amused by what they heard and saw—but, after all, they felt a bold spirit beating its wings in the American pioneers' tradition. Harvey W. Corbett, the chairman of the New York Architectural League, who introduced Fuller in July 1929, did not hide his admiration.[11]

A few years later, a color rendering in *Fortune* magazine pictured the Dymaxion house and the Dymaxion twelve-deck tower in a homogeneous ensemble of a newly built neighbourhood.[12] Fuller's unsophisticated sketches had been integrated by the hand of the *Fortune* illustrator into the discourse of a more conventional dwelling perspective. An image like this looks similar to analogous drawings of modern architects. Yet there is a fundamental difference between this and a dwelling project of Neutra, Le Corbusier, Gropius, or Oud. Their proposals always referred to the factuality of the city, the *civitas*, which was crucial for maintaining societal order. To Fuller the city plays no evident role as the core of civilization. He used large cities during his lifetime as sounding boards for his ideas: Chicago, New York City when he was young, Philadelphia in his later years. These cities were his social platforms; they supplied him with his audiences, and they nourished his imagination. The city for Fuller was a place where he could accumulate energy impulses to launch himself. Practically, he probably enjoyed living in the cities, but intellectually he made the city responsible for many diseases of society. His design initiative did not deal with the cultural phenomenon of the city at all.

Fuller often spoke about the problems of the city. He considered national economies to be egotistical; thus, the city—the condensed essence of a national economy—was actually on the wrong side. Fuller believed that the world needed a supranational structure and a global responsibility network; instead, nations remained attached to the erroneous image of a flat Earth and remained embroiled in an endless chain of armed conflicts about the exclusive usufruct of territories.

A rough and awkward sketch drawn on a sheet of letterhead, not dated, but certainly from 1928, indicates his image: a view of the 4D-multideck towers spread over the whole dry land archipelago of the globe. There are no nations, not even continents any more. The territory is Earth's dry land surface, a natural soil that can nourish the dwelling and building activity of all humanity around the globe.

Using this image of 4D towers as bases, Fuller refers, instead of to the "city," to another basic constant in American history: the settlers' experience of the "discovery" of unknown territories. He introduces the "lighthouse" as a building reference for the 4D house, akin in its geometry and in its attitude in marking a vanguard spot. The lighthouse is a sort of watchtower that guides the intrepid settlers who are making their way through the wild nature.[13] The Dymaxion house concept, although radically new, seems to be an offspring of a genuinely American experience of dwelling, one that is not marked by the city but by westward movement into the vast, empty prairies. Whereas in Europe a house is built into an existing frame of society, with a given organization and laws, in the history of the United States a house was, after all, an outpost.[14] There is a direct link between the American homestead, a lone outpost in the wilderness, and R. Buckminster Fuller's multideck tower and mooring-mast structure.

In his lectures on the 4D/Dymaxion house, and in the documented newsreels of his presentation, Fuller explains how the house will provide "shelter": "It must be proof against earthquakes, floods, tornadoes, cyclones, marauders, electrical storms."[15] Thus he introduces this elementary meaning of the home, and the shelter aspect has to do with realities of life in rural America, where people are constantly exposed to the perils of nature. Indeed, nature there is more threatening than in other regions of the world, since North America offers no such north-south weather-protection barrier as the Alps or the Himalaya.[16]

"The Anglo-Saxon origin of the synonym 'shelter' would be: SHELL scyld (shield) TER -trum (firm): That which covers or shields from exposure or danger; a place of safety, refuge or retreat."[17] Fuller himself, in biographical statements, credited the Dymaxion house with providing such a shelter and promising a convenient and safe environment to live in. Fuller and his wife, Anne, actually suspected that the poor living conditions in their drafty Chicago apartment had contributed to the death of their first daughter, Alexandra, who died in 1922 at the age of four from spinal meningitis and paralysis. In many ways the 4D house grew out of the deep depression and shock that Fuller experienced after this family tragedy and after he had lost his Stockade job. Fuller recounted a mystical experience along the banks of Lake Michigan that brought him back from the edge of self-destruction in 1927, and he committed himself to doing something good for humankind. He saw a congruence between his family's tragedy and the poor living conditions of other people all around the world. Thus, his radical 4D housing concepts grew out of his desire to achieve some kind of salvation for humanity.

"Spirit House"

The revolutionary impact of Fuller's novel idea is expressed in an emotional sketch, certainly a key document, dated February 9, 1928. This sketch, one of the first in his exciting venture, depicts a central mast with dependent radiant bearing beams. He names the sheet "Spirit House" and expresses the importance of his discovery: "Spirit House—The new tool! Metal! Fibre Stress / utilizing dynamics + tensile strength/as yardarm/gravity/self tuning or plumbing."

The Stockade memorial album contains a photograph of a load test in the shop: a wall is being pressed until breakage. In conventional architecture weight and load tend to weaken a structure, whereas a tension-based structure stiffens when loaded. Thus he mentions among the factors of his house: "gravity" and "self tuning." With his own approach Fuller had stepped outside the common reference system by the heuristic method of inversion.

The 4D/Dymaxion house proposed numerous unique or highly unusual features:

- The lightweight mass-producible metal structure in analogy to the mass-produced cars;[18]
- The central supporting mast and the radial structure;
- Floor construction by piano wire and with pneumatic surfaces;
- The omnidirectionality of the house (the first sketch was based on a rectangular plan, but Fuller switched to the hexagon in the spring of 1928, and the circular plan came later, with the Wichita House);
- The organization of the plan with the machinery and appliances (assembly units) being used as partitions;
- The triangulation of the plan; the abandoning of the right angle both in plan and also in elevation (window elements);
- The equipment of the house with radio, TV, and communication facilities, providing the inhabitants with a full range of access to information;
- The abandonment of individual property ownership in favor of the service concept—housing is a service industry rather than an individual asset, in analogy to the telephone service concept.[19]

This program offered unconventional, amazing, sometimes even outlandish ideas. In addition, I assume that this revolutionary concept was outside of the domain of the acknowledged discourse in architecture on either side of the Atlantic Ocean. I have already mentioned the correlation between the 4D-house concept and the settlers' movement in American history. We must not forget to mention the limits of this reference, or better, the point where it detaches from this tradition. To Fuller, contrary to the settlers' approach, the building ground was not a country, nor a nation, nor a personal claim. Fuller rejected the idea that land could be claimed as personal property; a territory was something to be used rather than possessed. Nor, however, was his view a socialist one, since property, private or collective, was not the point. He focused on use rather than exclusive or collective property ownership. He envisioned providing a housing service, something like telephone access at that time. Houses would be rented instead of owned, and housing would become a "world encompassing service industry" offering a reliable network of affordable, quality housing.[20] Cities would be supplanted by active transportation and communication between these bases.

One might assume that the 4D/Dymaxion house was nothing more than a relentless futuristic dream where the wonders of technology would overcome the old ways and means of housing. But this project was not based on technology or innovation for its own sake; the "Spirit House" was technology directed toward a humanistic goal: "Children are born truthful. They only learn deception, falsehood and instinct from the selfish prohibition of truth by their elders. . . . Let us solve the problem of the home, the housing of childhood, the prime reason for the home, and we will remove the majority of the traces of the dark ages of selfish unenlightenment."[21]

Words like these, written in 1928, may surprise anyone who thinks in terms of a thoroughly technical and materialistic reality, as was pushed forward in Europe after World War I. In France, during the 1920s, the meaning of "new spirit" was intentionally profane. Le Corbusier's magazine *L'Esprit nouveau* (new spirit) was the fanfare of an aesthetic and artistic confession and offered a modern lifestyle built on the conviction of its rational premises. Le Corbusier's *pure création de l'esprit* refers to artistic autonomy of composition and to integral rationality. "Nous doter d'une fenêtre mécanique! Nous architectes, nous nous contenterons fort bien avec un module fixe. Avec ce module, nous composerons," insisted Le Corbusier.[22] [Please provide us with a mechanical window element! We, the architects, will be most happy with such a module. In applying it, we will compose.] In these words we hear an artist on his quest for aesthetic intensity using mechanically produced building blocks. Fuller, however, did not search aesthetics for inspiration—he was responsive to natural sciences. His title "Spirit House" has, therefore, another connotation.

The Dedicated Management of the Home

Fuller's concepts of shelter differed from both the European image of a vanguard architecture and the image of American-inspired rationalization and mechanization. His project had spiritual roots in America itself; indeed, in this respect Fuller does not seem to be the "random element" that he claimed himself to be. Perhaps there is a special American method that allows technology and spirituality to be alloyed rather than mutually opposed; this alloy has evolved in due course in American history.

"In the following drawings are presented modes of economizing time, labor, and expense by the close packing of conveniences. By such methods, small and economical houses can be made to secure most of the comforts and many of the refinements of large and expensive ones."[23] These words could be Fuller's, but they are not. They come from Catharine and Harriet Beecher's remarkable book *The American Woman's Home* (1869), a rich compendium dedicated to "the Christian family." The Beecher sisters and Fuller have amazingly comparable views and a similar content; only the layers are switched. To the Beechers the argument is ethical (they want people to live so as to please God), and the means are technical; to Fuller the argument is technical, but the goal is ethical (Fuller wants people to succeed in their unique existence on Earth).

"The grand art of ventilating houses is by some method that will empty rooms of the vitiated air and bring in a supply of pure air by small and imperceptible currents."[24] Such a claim stunningly foreshadows Fuller's programs for the Dymaxion house and his later Dymaxion dwelling machine ("Wichita House") project of 1946.[25] Again the words are from the Beechers' book.

"The first and most indispensable requisite for health is pure air, both day and night. . . . There are two modes of nourishing the body, one is by food and the other by air,"[26] wrote the Beechers. Here, we can see them move with great ease between physical, physiological, and metaphysical arguments. They trace back the most important (metaphysics) to everyday living conditions:

> The human race in its infancy was placed in a mild and genial clime, where each family dwelt in tents, and breathed, both day and night, the pure air of heaven. . . . No other gift of God, so precious, so inspiring is treated with such utter irreverence and contempt in the calculations of us mortals as this same air of heaven. A sermon of oxygen, if we had a preacher who understood the subject, might do more to repress sin than the most orthodox discourse to show when and how and why sin came.[27]

This is a remarkable view, depicting the pragmatic, creative, and genuine attitude of the nineteenth-century Americans. It is remarkable with respect to Fuller, as well, because of the amazing correlation between the spiritual initiative of the Beechers (with a technical background) in 1869 and Fuller's spirit-based high-tech home of 1928.

In his programmatic "balance" comparison between the conventional house and the lightweight, lightful 4D tower, Fuller points out the progress of his conception: The conventional house is "tied up to [the] city sewerage system," whereas his house is "completely independent . . . , all in air—above dust area etc." Isn't this analogous to the Beechers, the dream of transcendence of technique into a sublimation of the human spirit, perhaps even a kind of ascension like in Mondrian's 1910 theosophical triptych "Evolution"?

The pure air that Catharine and Harriet Beecher praised obviously exceeded the operation of just "opening the windows." They conceived of a scheme of air currents from inside the house, from the staircase / chimney block—the same core of a house that would be emphasized in the early work of Frank Lloyd Wright some forty years later. The principle of inversion mentioned above with Fuller's shift from the "push" to the "pull" principle is evident already here. Fresh air comes from the center zone of the house, which is not identical with inside, since it is a part of "outside," however located in the center. The hypothesis cannot be worked out here in detail, but technical progress manifests itself often in the mastering of a problem by the principle of inversion. Rain could be drained internally in hollow columns in Joseph Paxton's Crystal Palace (1851), or—closer to Beecher and Fuller—the steam engine in the belly of a vessel had imperatively to be provided with oxygen by internal air ducts. And in 1924 Albert Kahn designed the Ford Laboratory Building in Dearborn, Michigan, with an air-conditioning system effected by hollow columns again. This building was widely admired for the brightness, cleanliness, and precision of its interior.

"The new building will be full of light (health giving light), Lithesomeness, and beauty hitherto unconceived of."[28] This is R. Buckminster Fuller's twentieth-century vision, to be arrived at by the application of airplane technology that would be used to build the prototype Wichita House in the 1940s. He was inspired by advanced nineteenth-century thinkers, indirectly (or perhaps directly) by Catharine and Harriet Beecher, and quite clearly by the New England transcendentalists, among them mainly by his great aunt, Margaret Fuller (1810–50), whom he discovered while he was in full pursuit of his vision. Fuller cannot be rightly contextualized without recognizing some distinctly American preconditions and inspirations. Only by recognizing how he was able to crossbreed technology and salvation, to blend the new architecture inspired by the

European initiatives with American individualism and ingenuity, can we pay proper tribute to Fuller's originality. He straddled the technical and spiritual worlds, insisting on the need to harness science to aid humanity. Fuller would often repeat his hope that the high technology (often developed initially for the military) would eventually be focused away from destruction toward construction, or "from 'killingry' to 'livingry.'"[29] Thus, in the earlier part of his career, he addressed himself to army generals—who provided a testing ground for his ideas, such as the Wichita House. But a few decades later, as he continued to speak about the need for adequate housing around the world, the hippie generation, too, could accept him as one of their own.

6 Energy in the Thought and Design of R. Buckminster Fuller

David E. Nye

"Are you going to hear Bucky?" The shaggy undergraduate who asked me this in 1972 was the last person who would normally attend guest lectures. However, during the flood tide of the counterculture at the University of Minnesota, where I was then a PhD student, R. Buckminster Fuller had achieved iconic status. About a thousand people went to hear him speak in the largest lecture hall on campus. This scene emphasizes that a large number of people responded to "Bucky's" ideas in the last decades of his life, and found in them an inspirational link between the humanities and design, between the counterculture and engineering. The symbols of that connection were the geodesic domes built by some communes and erected on many campuses as sturdy, usually temporary, shelters for various purposes. The undergraduates who flocked to hear him may have come away convinced that he was a fountain of original ideas. In retrospect, some, though by no means all, of this thinking can be traced to iconoclastic traditions in American thought and technological design. Yet there is no denying the forcefulness of R. Buckminster Fuller's public speaking, which was filled with apt expressions and striking images.

The task here, however, is not to explore Fuller's manifold relations to other designers or to the counterculture but to focus on the central place of energy in his life's work. I will consider this in two sections, the first dealing with Fuller's ideas about energy as expressed in public statements and publications, the second showing how these ideas were manifest in specific designs and projects. In his own life, of course, there was no such neat separation between thought and action.

I

Fuller's ideas about energy had great resonance during the oil shortages of the 1970s, when it seemed possible that the world had entered a long-term crisis and energy was a central topic in public debate.[1] With the price of gasoline soaring, in the midst of the new ecology movement that had launched the first Earth Day in 1970, Fuller's concept of "Spaceship Earth" had tremendous resonance with the young. In 1969 they had learned from the Apollo lunar landing to see their world as a fragile, beautiful orb televised from the surface of the moon. The counterculture was receptive to mavericks, to idealists, and to anyone who suggested ways to decouple from massive centralized systems of power. Fuller filled the void created when the young rejected the ideas of their parents and other authorities. He too rejected the status quo. He spoke frankly and openly of considering suicide in 1927, instead giving himself over to a life project to invent "energy-effective environmental-controlling artifacts that did ever more environment-controlling with ever less pounds of materials, ergs of energy and minutes of time per each realized functioning."[2] He wanted to make the world a safer and more energy-efficient place, and he framed his feasible design ideas in a larger system of thought that was organic, egalitarian, and democratic. Fuller also spoke of the present as a time of crisis, when humanity as a whole was taking a "final examination" in which Nature would discover if human beings would succeed. Would they use their intelligence on military weaponry and other ill-considered projects or on what he called "livingry," the technologies that allowed people to live in efficient comfort?[3] All these factors fed the enthusiasm of the thousands of eager listeners who heard his nonstop speeches at hundreds of venues and dozens of radio stations during the 1970s.

If Fuller's energy ideas found a particularly receptive audience during the fuel crises of the 1970s, they emerged well before then and certainly were not formulated in response to the shortages of that time. Fuller was embraced by the hippies and radicals of the counterculture, who learned how to build a geodesic dome from the *Whole Earth Catalog*, but he emerged from a tradition of engineering and industrial design that was prominent in the 1920s and 1930s, and whose hallmarks were modernism, streamlining, and the idea that "form follows function." Like Le Corbusier, Fuller built a compact and functional Dymaxion house that was "a machine for living in." Like the early airplane designers, he was inspired by the ideal of streamlined objects built from aluminum and other lightweight materials that moved almost effortlessly through space and whose fundamental shapes mimicked natural forms.

Fuller's views of energy also were rooted in an understanding of entropy as described in the second law of thermodynamics, formulated in the mid-nineteenth century and widely accepted by scientists by the time he was born in Massachusetts in 1895. They feared not global warming but the "heat death of the universe" as the sun inexorably cooled down, and they worried about the rapid depletion of forests.[4] In 1900 the inventor Nicola Tesla summarized Lord Kelvin's widely accepted view that human life on Earth was limited:

> From an incandescent mass we have originated, and into a frozen mass we shall turn. Merciless is the law of nature, and rapidly and irresistibly we are drawn to our doom. Lord Kelvin, in his profound meditations, allows us only a short span of life, something like six million years, after which time the sun's bright light will have ceased to shine, and its life-giving heat will have ebbed away, and our own earth will be a lump of ice, hurrying on through eternal night.[5]

From this perspective, science's urgent role was to prevent human beings from squandering the energy supplies available. Fuller certainly retained that idea, but he placed it in a new context.

By the time Fuller was ten, in 1905, the Newtonian world order was breaking down, as Einstein published his groundbreaking papers on relativity. Einstein's theory was becoming widely known just as Fuller reached adulthood, and much of his life's work could be seen as an attempt to see what $E=mc^2$ meant for the designer. In *Synergetics* Fuller developed a post-Euclidian view of the world, in which lines by definition cannot be straight but rather are "energy-event traceries, mappings, trajectories." Fuller concluded that "physics has found no straight lines: only waves consisting of frequencies of directional inflections in respect to duration of experience."[6] Likewise, the apparently simple and unproblematic idea of a "point" had to be rethought, and "the phenomena accommodated by the packaged word *point* will always prove to be a focal center of differentiating events." For Fuller, geometry did not describe a timeless space of pure Cartesian form but rather a universe where lines "cannot go through the same point at the same time."[7] Once one thought in these terms, building and designing on Euclidian principles became nonsense. A square house made with all right angles was an inherently inefficient form that mimicked a false geometry, while a geodesic dome, for reasons I will return to, embodied a more accurate understanding of the universe.

During Fuller's many talks and radio interviews of the 1970s, however, only the few who read his dense 870-page *Synergetics* could see the full complexity of his thought. As he had put it in the epilogue to *Utopia or Oblivion*, "The environment always consists of energy—energy as matter, energy as radiation, energy as gravity, and energy as 'events.'"[8] The general public paid attention to his ideas about efficiency, design, and energy use. Such ideas expressed during that visit to Minnesota also cropped up in *Critical Path*, notably, the argument that "Earthians" should be "able to live entirely within its cosmic-energy income instead of spending its cosmic-energy savings account (i.e. fossil fuels) or spending its cosmic-capital plant and equipment account (i.e. atomic energy)," which Fuller compared to "burning your house down in order to keep the family warm."[9] Fuller tirelessly proclaimed that this was completely unnecessary. As he declared in a keynote address at a conference on energy and the future of American communities: "There is no energy shortage. There is no energy crisis. There is a crisis of ignorance." He was convinced that "using only the technology available already, we can produce enough energy for everybody in the world, while phasing out all fossil fuels and atomic energy." With characteristic optimism he declared that "it is possible for all humanity to survive at higher standards than any have ever known while employing technologies that do no damage to the ecologically regenerative balance of the environment."[10]

Indeed, in retrospect one might see his life's work as an attempt to prove that "it is possible for all humanity to prosper while employing only the natural energy income of wind, tide, sun, gravity as water power, and electromagnetics of temperature differentials."[11] Fuller believed that recent history was the story of a "self-accelerating doing-more-with-less invention revolution" that could be exemplified by his own geodesic domes. They showed how much more efficient human beings could be. The "world's prime, vital problem (to which we must apply design science) is: how to triple swiftly, safely, and satisfyingly, the overall performance realizations per pound, per kilowatt, and manhour."[12] He estimated that the average machine was only 4 percent efficient, thus leaving enormous room for improvements, while in the average building he found "less than 1% overall structural efficiency," which meant that "we could build one hundred comparably volumed and useful buildings out of the same weight-, time-, and energy resource units now ignorantly processed into one building."[13] He was confident that the "normal rate of inventive evolution" would lead to a tripling of efficiency, with more comfortable and better lives for all. He called this trend of doing more with less, "ephemeralization,"[14] and it can be seen as the counterforce to entropy. As one of Fuller's oldest friends summarized, "the law of entropy may be a foundational building block of physics, but not for the human mind. Indeed, he would remind us, as had Thomas Huxley, father of Julian Huxley, nearly a century before, that the human mind can reverse that law. It is regenerative, not only bringing order out of chaos, inventive creativity, or in Bucky's trenchant phrase, 'Doing more with less.'"[15]

Fuller saw energy as inseparable from the environment. At an international conference in Reykjavik in 1977 he emphasized that "the environment . . . must really be thought of as not things, not scenery, but environment as the energy [of] both the metaphysical and the physical universe around us. The metaphysical environment is a most powerful one, the conceptioning that human beings have of their explanations of their experience."[16] Because he insisted on seeing energy not as fuel (or as an isolated thing in itself) but as part of a larger system, Fuller used the term *synergy*: "Synergy is to energy as 'whole' is to 'part.' Synergy is to energy as integration is to differentiation. Energy studies separate out—isolating particular phenomena out of the total phenomena of Nature. . . . Synergy is the associate behavior of wholes within Nature." Fuller resisted the compartmentalization and specialization of science, which dissected phenomena but often did not put them together again. On the abstract level, Fuller concluded in *Synergetics*, "nature uses the tetrahedron as the prime unit of energy, as its energy quantum, because it is three times as efficient in every energetic aspect as its nearest, symmetrical, volumetric competitor, the cube."[17] To enclose space, therefore, the triangular form was quite literally the natural alternative when Fuller turned to the practical level of building the geodesic dome.

Both Fuller's integrative thinking and his penchant for reasoning from nature go back to transcendentalism. A great nephew of the transcendentalist Margaret Fuller, he celebrated that connection, lamenting that, compared to Emerson, she had been forgotten, an oversight that has since been corrected.[18] Like his great aunt, as well as Emerson and Thoreau, Fuller thought in holistic terms and refused to see the world as the mechanistic assemblage of its constituent elements. Rather, it was an organic whole that was greater than the sum of its parts. Fuller also inherited transcendentalism's iconoclasm, self-reliance, and aesthetics. The sculptor Horatio Greenough (1805–52) was that movement's seminal artistic figure. He rejected imitation of the past and praised simplicity and efficiency in design, values that would later emerge in Fuller's work.[19] Like many other designers and architects influenced by transcendentalism, including Henry Ford, Louis Sullivan, and Frank Lloyd Wright, Fuller prized functionality over surface decoration. The buildings of Sullivan and Wright (both widely discussed in Fuller's early decades as exemplary American structures) broke with European architecture, adopted new building materials, and proclaimed that form should follow function. Indeed, in later life Fuller and Wright became friends.[20] In short, Fuller's conceptions of energy and design flow from iconoclastic impulses (one hesitates to call iconoclasm a tradition) that have long been encouraged and justified by transcendentalism.

II

Having seen how thoroughly energy (or synergy) was infused into many aspects of Fuller's thought, including his redefinition of the geometrical point and line, as well as his design work and his displeasure with the extensive military buildup, it is time to look at the place of energy in the projects he either built or proposed. Fuller believed that the process of electrification was a fundamental sociotechnical transformation of human relations and called for the erection of a globe-spanning power network. As he put it in *Critical Path*: "The development of our *omni-world-integrating electrical-energy network grid* which will realistically put all humanity on the same economic accounting system and will integrate the world's economic interests and value systems and lead most swiftly to the realistic elimination of the 150 sovereign-nation systems, needs only a relatively few geographical interlinking operations. It does not need the invention and development of new technologies."[21] Fuller's proposed electrical grid would circle South America, link it with North America, and cross the Bering Strait from Alaska to the Soviet Union, and from there cross Asia to Europe, and then swing down to Africa. This "Global Energy Network International" (GENI) would make the most efficient use of generating capacity, sending surpluses in one part of the world to satisfy demands elsewhere. Rather than a balkanized system of local power plants, where every community built capacity well beyond the average demand in order to deal with peak demand, a world-spanning system would not have sharp peaks in average demand, as it smoothly transferred electricity wherever needed. Note, too, that Fuller expected the construction of such a system to weaken nationalism. The famous General Electric scientist Charles Steinmetz had argued in the 1910s that full electrification would force societies to evolve away from competitive capitalism to cooperative socialism.[22] However, the long-distance transmission capabilities of that time made a world electrical grid impossible. But in the second half of the twentieth century, long-distance power transmission technologies more than doubled the distance they could cover, and the reasons such a system were not built became increasingly financial and political. Both Fuller and Steinmetz expected a universal electrical grid to undermine nationalism and to teach human beings that they were interconnected. An institute still devoted to realizing this goal credits Fuller with the idea and explains the fundamental idea on its Web site: "All the earth's resources were catalogued, and human survival needs were assessed, giving world planners the potential for global thinking and solutions. Upon realizing that electricity was the common denominator of all societal infra-systems: food, shelter, health care, sewage, transportation, communication, education, finance—the priority of delivering sufficient power to every human was established. Access to electricity for everyone is a primary measure of a modern society."[23]

Like Marshall McLuhan, also widely influential in the 1960s and 1970s with his idea that changes in media shape changes in society, Fuller at times seemed to believe in a form of technological determinism, in which social change would automatically flow from alterations in the infrastructure. Fuller, however, was hardly a determinist in practice. National governments did not rush to adopt GENI or most of his other ideas. He knew that many people and institutions resisted unfamiliar designs and that energy efficiency was not automatically adopted. Therefore, he called on his audiences to get involved in what is today called "the social construction of technology." For example, at the University of Ohio he inspired and advised a group of students interested in windmill design, who worked for several years under the direction of a faculty member.[24] On a larger scale, he instituted a "World Game," which annually took place on a college campus, including the University of Massachusetts and the University of Pennsylvania. Unlike the conventional conference where a few speakers lecture and most people listen, the World Game involved all participants interactively in thinking and planning for the future.

Fuller had planned for the future all his life. When he came of age, the world of manufacturing was in the throes of rapid change. By the time Fuller was twenty, in 1915, Henry Ford had amazed the industrial world with his assembly line, which literally drew crowds at the San Francisco Panama Pacific Exposition, while industrial tourists kept a permanent staff busy showing off the Ford factories in Detroit.[25] The budding inventor soon sought to apply Ford's ideas to architecture, in the form of mass-produced, mobile housing. This interest found full expression in his 4D house. In 1927 he mimeographed two hundred copies of the design and circulated them as a call for inexpensive mass-produced housing. The idea caught the eye of executives at Marshall Field & Co., who asked him to build a scale model for display at the company's main store in Chicago. It attracted interest but not investors. Lacking funds to build a prototype, Fuller turned to other tasks for a decade, notably the Dymaxion car. Its streamlined, teardrop design was not uniquely his own, as others worked in the same vein, notably Norman Bel Geddes, who also visualized and built such automobiles. Indeed, because its form promised energy efficiency and speed, the previous year the Society of Automotive Engineers had endorsed the shape as "the final evolution" of the automobile's design.[26] Yet Fuller did make one of the few working prototypes, and it brought him recognition.

During World War II Fuller returned to mass-produced housing, producing structures for the military. These circular buildings prefigured the more thoroughly worked-out Dymaxion house, which he completed in the mid-1940s. A prefabricated structure assembled from standardized parts, its sleek, metal design was characteristic of the streamlining of the 1930s. Fuller intended a mass-produced, affordable house that was transportable and environmentally efficient. It would enable the owner to move more easily and to use a smaller amount of energy than in a conventional home. To mass-produce it, Fuller turned to Beech, an aircraft manufacturer accustomed to fabricating with aluminum and other high-tech materials. It was to be sold "for the price of a Cadillac, and could be shipped worldwide in its own metal tube."[27] However, conflicts and disagreements about how to translate his design into a manufactured product derailed the project.[28]

Fuller continued to conceive of housing in terms of energy, though not merely in terms of being heat-efficient or streamlined. Two decades later, he declared:

> Thinking correctly of all housing as machinery we began to realize the complete continuity of interrelationship of such technological evolution as that of the home bedroom into the railway sleeping car, into the automobile with seat to bed conversions, into the filling station toilets, which are accessories of the parlor on wheels. . . . All this living machinery complements the inherently transient nature of world society and its progressive emancipation from the local shackles of physical-property "machines" which were so inefficient and so enormous.[29]

In this passage Fuller moves well beyond the usual modernist conception of the house as a machine for living, to imagine a genealogy of machines rapidly evolving from the conventional house (conceived as an immobile prison that trapped its owner) toward compact, mobile systems of amenities. He concluded "that the transition to the faster technologies, which will open up all oceans and skies to man's support and enjoyment, is an inevitable consequence of what is already irrevocably and inexorably underway." And what was that? "The comprehensive introduction of automation everywhere around the earth will free man from being an automaton and will generate so fast a mastery and multiplication of energy wealth by humanity that we will be able to support all of humanity in ever greater physical and economic success anywhere around his little space ship Earth."[30]

This was not merely a rhetorical vision. Fuller designed mobile structures to allow human beings to move about more easily. Noting that the average American family moved frequently, he created structures that they could take with them. Using some features known from yacht and mobile-home design, and adding many more of his own, he wanted to provide more space than a trailer contained but less than a typical home. These structures recur throughout his design work, from the late 1920s onward. An example from late in life was the "Fly's Eye Dome." It was twenty-six feet in diameter and large enough for two floors. Mass-produced from lightweight hard plastic, unlike the geodesic dome, it did not need to be assembled onsite but could be moved from place to place by helicopter and provide nomadic shelter for the peripatetic American family. Energy demands for heating or cooling would be kept low by double walls. Most important, the "Fly's Eye Dome" did not require a link to local utilities, as it could harvest energy from the sun and wind, gather rainwater in a cistern, and recycle waste to produce methane gas. "The basic hardware components will produce a beautiful, fully equipped, air-deliverable house that weighs and costs about as much as a good automobile."[31] Given the rising costs of housing, something akin to this vision might yet emerge.

By far the most famous structure that Fuller created is the geodesic dome,[32] which since 1950 has been erected on every continent and used for a wide variety of purposes. The relationship between the domes and energy is not limited to their design but also finds expression in their function. The domes were conceived as "environmental valves, differentiating human ecological patterns from all other patterns."[33] This was obvious in the case of the domes erected in the Arctic as part of the Defense Early Warning system, as these fifty-five-foot-diameter structures kept out the cold and wind, making it possible for the remote radar stations to function. Similarly, every dome operates as a valve between an inside and outside. Because geodesic domes are made from identical parts, they can be quickly assembled, usually in fewer than twenty-four hours. This is not only energy-efficient building; it also enables construction in severe climates or adverse weather conditions where slower conventional methods would render a project difficult or impossible. When the structure is completed, one can see from the outside a series of interlocking triangular units that form a globular structure. The individual units appear flat from a distance, but just as Fuller argued there are no straight lines in contemporary physics, each strut is slightly curved to fit the overall arc of the particular sphere. When all the individual triangular units are linked together, the resulting dome spreads out the stresses of the structure, so that it is evenly distributed over the whole surface.

If all domes are in principle identical in conception, however, size does matter. Larger-scale domes are usually more satisfactory to people who must spend time inside them. In a small one the sides slope upward so sharply that the upper half is not particularly useful space. At ground level, curving walls made it hard to place furniture on the sides. However, as a dome grows in diameter, the curvature of any small section of wall is far less pronounced. Furthermore, in large domes the space can be easily carved up into a series of internal levels. Thus, the larger domes obviate many of the practical problems of the smaller-scale units, while at the same time becoming stronger as wind shear and other load factors are distributed to all the elements. That is why Fuller could plausibly imagine parts of cities covered by domes, reducing their energy consumption for heating or cooling.

Fuller early on proposed a gigantic dome two miles in diameter and one mile high to be constructed on the northern end of Manhattan (fig. 6.1). This was not worked out into a full design, but in 1971 he created a far more detailed plan, as chief architect of the Old Man River Project, a domed city one-mile in diameter that was to transform the largely poor and black neighborhoods of East St. Louis (fig. 6.2). The Old Man River Project was never built, perhaps in good part because it was not merely a dome over an existing city but a far more costly and visionary idea: an entirely new structure organized with the conviction that cities had to be rethought from the ground up. As Fuller put it, "Cities developed entirely before the thought of electricity or automobiles or before any of the millions of inventions registered in the United States Patent Office. For eminently mobile man, cities have become obsolete." It was necessary to rebuild, to demolish the old buildings and replace streets, water lines, and sewer

Figure 6.1
Sketch of proposed dome over Manhattan.
 Source: Special Collections, Stanford University Libraries.

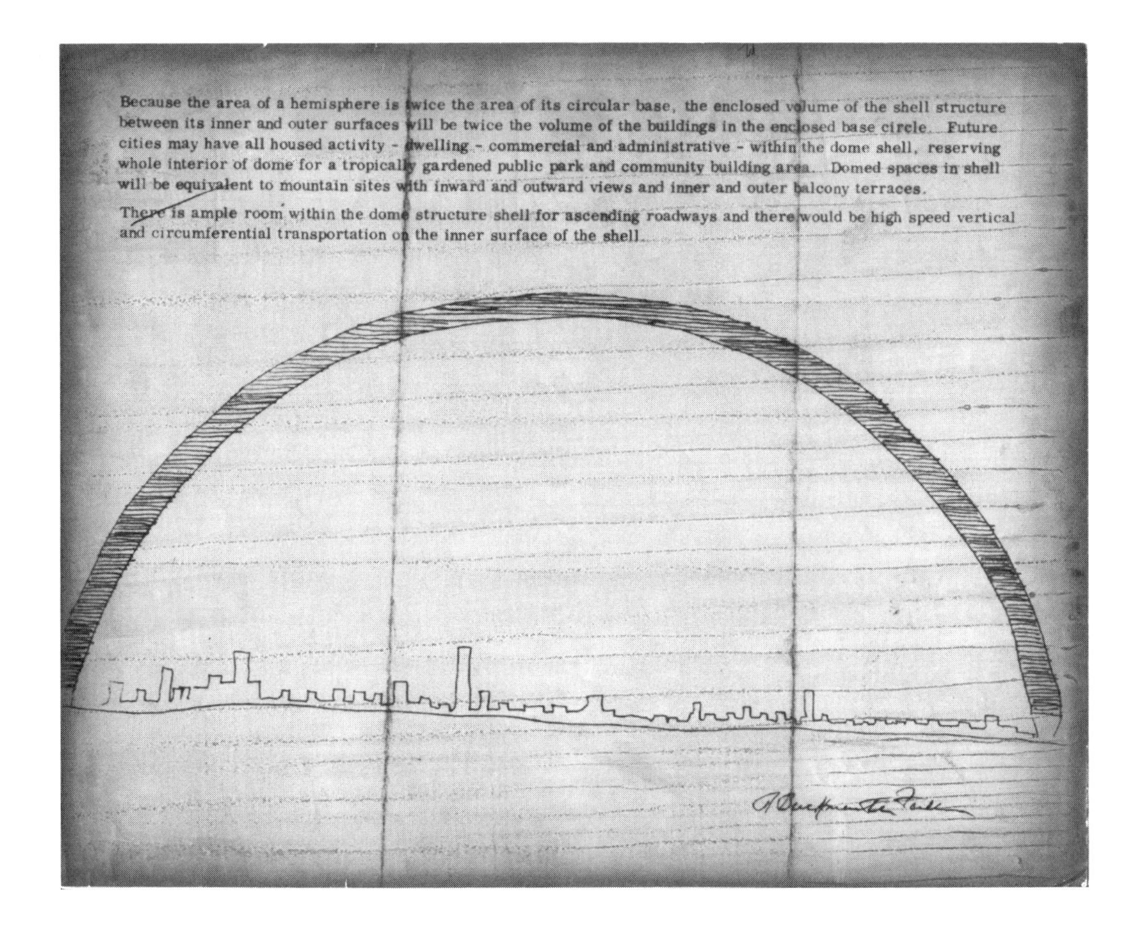

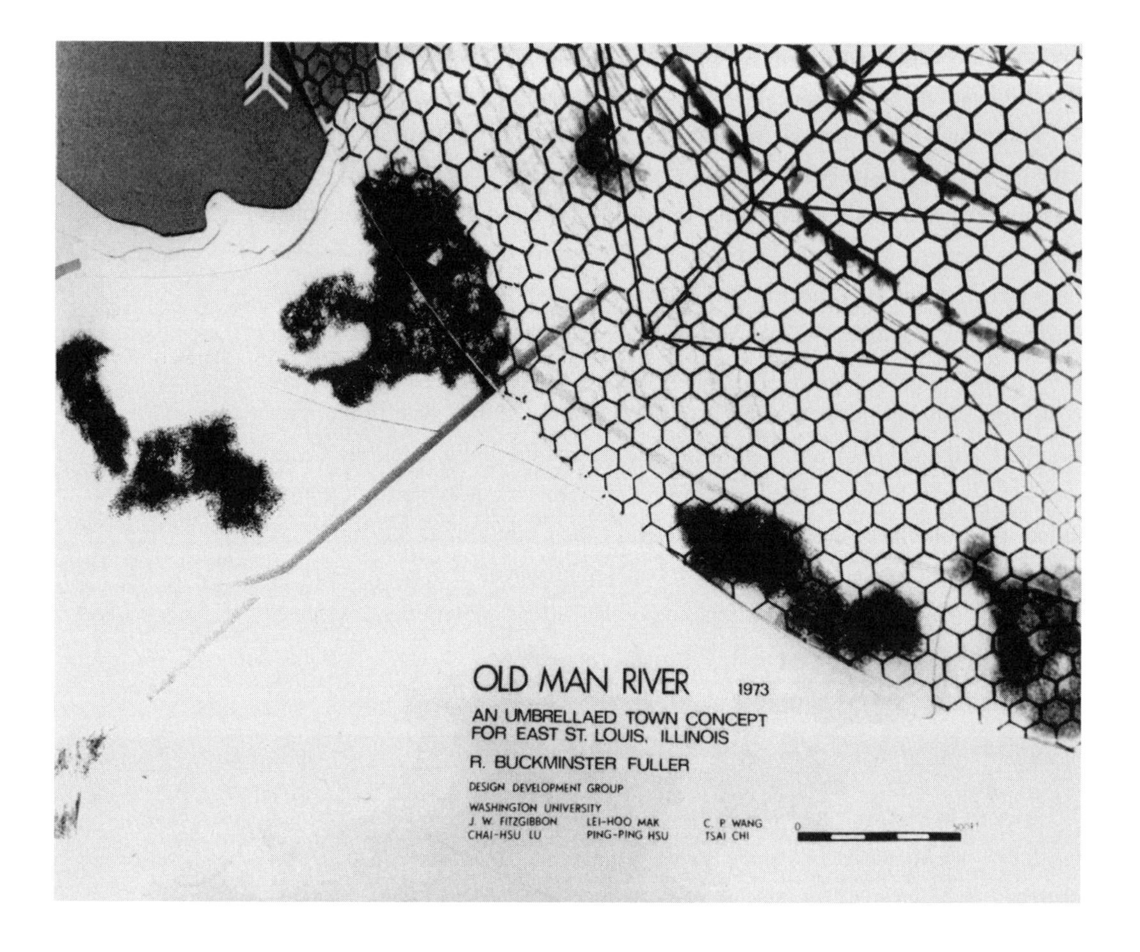

Figure 6.2
Cover of Old Man River proposal for East St. Louis (1973). Source: Special Collections, Stanford University Libraries.

lines, and to give up on "yesterday's no longer logical overall planning geometries." The proposed design appeared something like a moon crater and consisted of a circular building with fifty curved terraces. The inward-facing terraces contained stores, offices, tennis courts, athletic fields, and all the amenities of public life. On the outside of the circle structure, facing outward and offering more privacy and stunning views, were thousands of apartments, divided from one another by hedges and gardens (fig. 6.3).[34] Conceivably, such a collectivized living arrangement might have failed the sociological test of having thousands of families live in it, but there can be no doubt that its shared walls, shielded from the winter's cold, would have been far more energy efficient than individual houses. Likewise, the entire space could also have been air-conditioned at far less expense owing to economies of scale. Fuller explained:

> Throughout the year, Old Man River's City will have a naturally mild climate. With a large, aerodynamically articulated, wind-and-weather-controlled ventilator sys-

tem atop and round the dome, together with the 500-foot-high vertical opening that runs entirely around the city below the umbrella, the atmospheric controllability will guarantee fresh air as well as energy conservation. The umbrella will jut out above and beyond all the outer-slope residential terrace areas as does a grandstand roof, so that neither rain nor snow will drift horizontally inwardly.[35]

Just as important, the concentration of the population would have eliminated the need for automobiles, which had no place in this new urban center.

Though this gargantuan project was not attempted, geodesic domes were successfully adopted by many corporations, international agencies, and world's fairs, beginning in the early 1950s. Quite possibly the majority of the earth's population has seen a geodesic dome somewhere. Yet it does not seem to be fully understood. The most famous of his geodesic domes remains the United States Pavilion at Montreal's Expo 67. One architecture critic typically summarized it as "a giant dome, roughly three-quarters of a sphere, designed to look like a lacy filigree weightless against the sky. Height: 200 feet; spherical

Figure 6.3 Photograph of the model for the Old Man River proposal. © Estate of R. Buckminster Fuller. All rights reserved. Used by permission. Source: Special Collections, Stanford University Libraries.

diameter: 250 feet. Construction: a space frame of steel pipes enclosing 1,900 molded acrylic panels."[36] However, the geodesic dome is far more than a clever and attractive design that uses a small amount of material to enclose a large space and becomes stronger the larger it is made. The physical properties of the dome also manifest Fuller's synergetic thinking. Domes embodied his concept of "ephemeralization," or doing more with less, showing that energy efficiency is not only a matter of making incremental improvements in existing designs and techniques. The dome embodied radical new thinking, not only in its overall shape but also in the construction of its individual components. In *Synergetics* every line is understood as an inherently dynamic and always slightly curved element, and every object exists in time as well as in space. The geodesic dome actualized Fuller's ideas in a visible, functioning pattern. As thousands of domes went up in all parts of the world, they were "tangible, measurable illustrations of laws fundamental to the nature of the universe, of the spread and temper of energy patterns. . . . The domes perform according to the predictions of Energetic Geometry."[37]

Like the shaggy Minnesota undergraduate who enthusiastically went to hear Fuller in 1972, the tens of thousands who heard him speak during the last decades of his life probably had not read *Synergetics.* However, they could grasp the simplicity, strength, and promise of this new form of architecture. It was economical with materials and therefore environmentally friendly. Undercutting the specialization of knowledge, it was easy to assemble and could likewise be disassembled and moved. A geodesic dome also could be adapted to make the most of passive solar energy or wind power. From Fuller's talks, people learned to see the dome as part and parcel of a larger philosophy of energy that harked back to transcendentalism and that could be linked to other organic ways of seeing nature, while at the same time demonstrating its empirical validity as a working design. Fuller explained the energy crisis as a human failure to use resources intelligently. It was a failure that could be corrected and not a shortage of raw materials that spelled unavoidable hardship. When one looked at a geodesic dome, Fuller's concept of ephemeralization not only made sense, but it also seemed the inevitable way forward. Whether his listeners preferred the stand-alone self-reliance of the "Fly's Eye Dome," embraced the idea of a global energy network linking all nations in a single electrical distribution system, or wanted to enclose their community in a protective giant geodesic dome, Fuller offered striking and original solutions that still may inspire future developments. Listening to him, it seemed, indeed, that "there is no energy shortage. There is no energy crisis. There is a crisis of ignorance."

7

Necessary Beauty

Fuller's Sumptuary Aesthetic

Jonathan Massey

> Beauty rests on necessities. The line of beauty is the result of perfect economy. . . . There is not a particle to spare in natural structures. In rhetoric this ART OF OMISSION is the chief SECRET OF POWER.
>
> Ralph Waldo Emerson

Buckminster Fuller's many designs and inventions have long been celebrated for the ethical virtues of their efficient engineering and production. The aesthetic appeal of these designs, on the other hand, has often been considered a circumstantial feature—fortuitous confirmation of their fundamental rationality.[1] This perception, bolstered by Fuller's own insistence that aesthetic considerations were strictly secondary to efficiency criteria, has distorted our understanding of his work. Aesthetic strategies were essential to Fuller's pursuit of an ethically superior society through efficient design. The regular geometries that Fuller used to optimize the performance of structures and machines also lent them a distinctive beauty that convinced investors and consumers to finance and buy them and so to enlist in a voluntaristic social reform project. Fuller's exploitation of geometry for its rhetorical power is manifest in many of his key projects, including the geodesic domes for which he is best known. Their genesis lay in his 1928 development of the 4D house, better known as the Dymaxion house (fig. 7.1).

Fuller's use of regular geometry to construct sturdy and efficient structures was inspired by the practices of engineers, especially the efficiency engineers who followed Frederick Winslow Taylor in applying scientific management principles to industrial production and other social processes. Yet Fuller's use of geometry exceeded the requirements of structural and manufacturing efficiency, taking on a rhetorical role that was shaped by the work of architects, particularly that of Claude Bragdon, a modernist who in 1915 developed a system of universal ornament to integrate architecture, art, and design. Fuller synthesized aspects of these two distinct traditions through the conceptual rubric of the fourth dimension. He associated the Taylorist use of time controls in the manufacturing process with Einstein's theory of relativity and the concept of a temporal fourth dimension. By reducing the time needed for both the production of housing and its maintenance by occupants, Fuller incorporated "four-dimensional" time efficiencies into his designs. Bragdon's

Figure 7.1 One of Fuller's 4D-house models, ca. 1929. Source: Special Collections, Stanford University Libraries.

system of ornament, on the other hand, had been based on an older concept of a spatial fourth dimension that carried with it ethical imperatives for altruistic behavior. Projecting four-dimensional shapes into three- and two-dimensional patterns, Bragdon had enlisted the beauty of geometric order to regulate design and consumption in what he considered socially beneficial ways. By adapting the geometries with which Bragdon had expressed his four-dimensional ethical code and sumptuary ethos, Fuller incorporated Bragdon's rhetorical use of beauty into his housing reform project. By linking technocratic production efficiencies to aesthetic strategies for using geometry to modify consumption, Fuller was able to reconcile technocratic faith in a single best pattern of social organization with his libertarian commitment to individual self-determination.

Comprehensive Design

Over the course of several months in 1928, Fuller outlined a system of industrialized housing that promised to transform human society by releasing parents from unnecessary labor and giving children the benefit of an improved home environment. In May of that year he privately published these ideas in a manuscript, called *4D Time Lock*, that combined attributes of philosophical treatise, mystical statement, reform manifesto, and business prospectus. The *Time Lock* outlined Fuller's vision of a world integrated by increasingly efficient manufacturing and transportation systems that would free up time, energy, and material so that families could enjoy lives of nomadic leisure. The manuscript invited readers to invest in an association, called the 4D Control Syndicate, dedicated to manufacturing and renting prefabricated houses. Fuller designated these projects "4D" to mark the central role the fourth dimension played in his reform vision.[2]

As he wrote the *Time Lock*, Fuller was developing designs for a lightweight, centrally supported metal house suitable for mass production. His initial design featured a square structural core that contained plumbing, services, and ventilation equipment while also supporting rectangular cantilevered floors. During the next few months Fuller redesigned the core as a slender mast from which floors hung by tension cabling and gave the house a hexagonal floor plan on a triangular module. Fuller drew the house in many different configurations, including both a single-family version and multistory apartment dwellings. By September 1928, Fuller's single-family house design had crystallized into a hexagonal one-story structure, suspended above the ground on a central mast, anchored to the ground by metal cabling, and furnished with a roof deck. Airlifted from factory to building site by a zeppelin, its mast anchored in a bomb-crater excavation, this "autonomous dwelling unit" would be installed virtually anywhere that its nomadic family found the best opportunities for work and for leisure. In Fuller's imagination the house would liberate families from geographic constraints and local allegiances, enabling them to take the best advantage of the opportunities afforded by the global market for labor. Mobile dwellings would stabilize the economy by creating a self-regulating labor market in which workers followed jobs. Fuller began exhibiting and publishing this project, which he called the 4D house, in fall 1928. He refined it in drawings and models over the next several months, adding secondary features such as exterior louvers and built-in furnishings and appliances. In April 1929, renamed the Dymaxion house by adman Waldo Warren, Fuller's design was exhibited at the Marshall Field's department store in downtown Chicago and publicized widely in the press, launching Fuller's remarkable career as an inventor, designer, author, and educator.

The *Time Lock* laid the groundwork for Fuller's later books and lectures. Similarly, the 4D house contained the germ of most of Fuller's subsequent design work. The principles of lightweight metal construction, industrial mass production, tension cabling, and geometric ordering that characterized the 4D house informed designs for the Wichita House, geodesic domes, "geoscope" information displays, tensegrity structures, and many other projects. Fuller's transportation and cartographic designs, such as the Dymaxion car, Autonomous Wing, and Dymaxion air-ocean map, meanwhile, were tools of mobility for a nomadic global society. By using design to increase efficiency and rationalize the labor market, Fuller aspired to reduce waste, including both inefficient production and what he considered unnecessary consumption. Fuller was inspired primarily by the abundance and harmony that could result from maximizing the benefits of industry and distributing them equitably in what he called "the socialization of essentials and plenitudes."[3] This process would increase individuals' life expectancy and standard of living by providing better nutrition, housing, and recreation. In doing so, Fuller believed, it would help the human species as a whole avoid self-destruction by relieving the scarcity pressures that stimulated competition, class strife, and warfare.

The implementation strategy of Fuller's social reform project eschewed both autocratic solutions and electoral politics in favor of market-based approaches. Fuller proposed a worldwide technological evolution to be carried out by individuals and corporations working to design and market better machines and products. He came to call this practice "design science" and envisioned it being practiced by "'comprehensive designers' who would coordinate resources and technology on a world scale for the benefit of all mankind, and would constantly anticipate future needs while they found ever-better ways of providing more and more from less and less."[4] By progressively increasing the efficiency of human resource use, comprehensive designers would relieve the survival pressures placed on individuals and the species by resource limitations.

Self-discipline

Fuller's work falls within the tradition of sumptuary regulation, the regulation of consumption in the service of social and political goals. Since antiquity, sumptuary codes have maintained particular aspects of social order by guiding the choices individuals make in purchasing and displaying goods.[5] By identifying some desires as excessive or luxurious, and so as illegitimate, they have regulated expressions of private desire in the name of the public good. From the Middle Ages through the seventeenth century, sumptuary regulation was frequently enforced through laws that specified what individuals wore and ate, what furnishings they possessed, and how they conducted funerals and weddings. In modern society sumptuary regulation has more often operated through economic incentives, such as those embedded in tax code provisions, and aesthetic codes, such as those promoted by modernist architects, than through outright prohibition. By making the individual responsible for his or her own self-regulation, modern sumptuary codes have replaced the old externally imposed limitations with a set of internalized disciplines.[6]

Disciplining desires—his own and those of others—was a major preoccupation for Fuller, who characterized his entire career as a series of "self-disciplines" of broadening scope.[7] While Fuller's career, devoted to using design to bring individual desires into alignment with one another, had roots in both the Technocracy movement and Bragdon's projective ornament, its development was spurred by the tensions in his own personality between appetite and duty. Fuller's 1928 outpouring of innovation was catalyzed by a delayed reaction to the death of his first daughter, Alexandra, who had succumbed to polio in November 1922 after surviving earlier bouts of influenza and spinal meningitis. Fuller blamed Alexandra's death on World War I, which had diverted resources from life-enhancing purposes such as public health services, resulting in such calamities as the 1918 influenza epidemic that struck the infant Alexandra.[8] At the same time, Fuller blamed himself, feeling that his daughter's death had resulted in part from his frequent absences from home due not only to work obligations but also to his self-centered appetite for drinking, gambling, and other bachelor pleasures. The birth of a second daughter, Allegra, in 1927, reactivated in Fuller the emotions associated with Alexandra's death. His new fatherhood coincided with a personal economic crisis: his removal from leadership of the Stockade Corporation by the new majority shareholders when his father-in-law sold off shares in the company. Fired from a company he had helped to found and build, Fuller felt himself—and his newborn daughter—victimized again, this time not by war but by finance capitalism.[9]

Offered a second chance as husband and father, Fuller contemplated the possibility that he would be a better provider dead than alive. Facing these anxieties in an existential crisis on the shores of Lake Michigan in fall 1927, Fuller considered drowning his sorrows and feelings of guilt in the icy lake, leaving his wife and new daughter to collect insurance money and allowing Anne to take a more suitable second husband. As he contemplated suicide, however, Fuller experienced a revelation during which, he later recalled, time stopped, he levitated, and a disembodied voice addressed him. "You do not have the right to eliminate yourself," it said, proclaiming it Fuller's duty to devote his knowledge and ability to "the highest advantage of others."[10] This mystical encounter convinced Fuller to dedicate himself to serving humanity as experiment "Guinea Pig B," beginning with a term during which he vowed not to speak to anyone else. Channeling his frustration and self-hatred into an intense productivity enhanced by this monastic discipline, Fuller devoted himself to reforming society by eliminating the "chaos" caused by war and market capitalism, both of which denied men, women, and—most painfully—children the benefits of the scientific and industrial revolutions. Projecting his commitment to personal reform outward onto the society around him, Fuller set out to redeem himself by saving children through better design of housing and other amenities.[11] Fuller's passion for taming his own "bestial self" carried over into his work as a lifelong campaign to regulate the consumption of others by reorganizing the "mechanical arrangement" of society so that individual selfishness would make the individual "inadvertently selfish for everyone."[12]

Fuller's Fourth Dimension

Fuller frequently attributed the efficiency and elegance of his designs to their distillation and replication of principles evident in nature. While this claim recurred throughout his career, the specific terms in which Fuller understood nature—and with them, the ways he claimed to replicate natural principles in his designs—changed over time. In the 1970s, for instance, Fuller liked to illustrate his claim that geodesic domes were based on natural principles with drawings of radiolarians, single-celled oceanic protozoan organisms. While nineteenth-century biologists such as Ernst Haeckel had documented many types of radiolarians in a variety of shapes, Fuller selectively featured radiolarians that took the form of faceted spheres, like his geodesic structures. When promoting the 4D house in the late 1920s, however, Fuller had compared its mast to the trunk of a redwood tree and its pneumatic rubber floors to "the life cell," nature's "primary structural member . . . a globule in which elements in their liquid and gaseous states are compressively enclosed by elements in their solid and tensed state."[13] Fuller's "nature" was a moving target that mirrored the technology with which he happened to be working. It was fundamentally a rhetorical tool lending the authority of "necessity" to Fuller's designs.

The fourth dimension was a recurring theme throughout Fuller's career. As with his other concepts of nature, Fuller's understanding changed over time, oscillating between—and sometimes combining—differing concepts of the fourth dimension as space and as time.[14] This ambivalence as to whether the fourth dimension was spatial or temporal reflected his familiarity with two disparate concepts of the fourth dimension. The concept of the fourth dimension of space had gained prominence following the 1868 publication of

G. B. F. Riemann's theory of *n*-dimensional space.[15] By demonstrating mathematically that space could possess a variable and potentially infinite number of dimensions, Riemann suggested that the universe might contain spaces of more than three dimensions. After World War I the new concept of the fourth dimension as time, formulated by mathematician Hermann Minkowski in 1907 and incorporated into Einstein's General Theory of Relativity, eclipsed the idea of higher spatial dimensions.[16]

For a half century after its publication, Riemann's work inspired speculation as to the potential reality of higher-dimensional spaces, including a large parascientific literature that posited a fourth spatial dimension as the explanation for occult phenomena and mystical experiences. In the new genre of "hyperspace philosophy" Riemann's discovery became a vehicle for social critiques and religious convictions, ranging from the social commentary of E. A. Abbott's satire *Flatland: A Romance of Many Dimensions* (1884) to the mystical doctrines of theosophy, the "spiritual science" that sought to reconcile modern Western science with ancient Eastern religious principles from the Bhagavad-Gita and the Upanishads. According to most hyperspace theories, the fourth dimension was a real space beyond the range of normal human perception, awareness of which had existential, epistemological, and ethical consequences. In the late 1870s, for instance, Leipzig physicist and astronomer J. C. F. Zöllner developed a theory of "transcendental physics" that explained spiritualist phenomena, such as clairvoyance and the materializing of objects within sealed enclosures, as fourth-dimensional phenomena. From the early 1880s to his death in 1904, English mathematician Charles Howard Hinton made the principle of a four-dimensional intelligence looking down into an exposed third dimension the basis of an altruistic ethical code. Hinton encouraged his readers to cultivate four-dimensional vision so that they might "cast out the self" and achieve transcendent unity with higher-dimensional cosmic being. In Russia, P. D. Ouspensky's *Tertium Organum* (1911) drew on Zöllner, Hinton, and other sources to develop a mystical cosmology characterizing the evolution of consciousness as a conquest of successively higher spatial dimensions. In Rochester, New York, meanwhile, Claude Bragdon synthesized ideas from Hinton and other hyperspace sources in articles and books that identified the fourth dimension as the future home of perfected humanity. By overcoming their materialism and transcending their egotism, Bragdon argued, individuals could gain access to a four-dimensional New Jerusalem where millennial dreams of abundance and harmony would be fulfilled. Bragdon disseminated these ideas in his books *Man the Square* (1912), *A Primer of Higher Space* (1913), and *Four-Dimensional Vistas* (1916). When he translated and published *Tertium*

Organum in 1920, Bragdon also introduced Ouspensky's four-dimensional cosmology to English-speaking audiences, who devoured the book in new editions issued almost annually.

Fuller combined aspects of both the spatial and temporal fourth dimensions, especially in his early career. Although he associated the 4D house and *4D Time Lock* with Einsteinian relativity, Fuller fused this temporal fourth dimension with ideas and representational strategies from hyperspace philosophy. He understood Einstein's theory to describe a "time-rate world": a universe in which energy moved bodies of varying mass at differential rates of speed.[17] Fuller's grasp of Einstein's theory was hazy, as he admitted to his father-in-law.[18] For the most part, he filtered relativity through his intuitive familiarity with the interrelationship of mass, energy, and velocity acquired summering at the Fuller family's Bear Island, then serving in the Navy during World War I. In *Nine Chains to the Moon*, for instance, Fuller described Einsteinian relativity as "a concept of the universe, all parts of which are in constant motion, as powered by unit energy in relative rates of speed of motion proportional to the frictional relationships of all the parts."[19] Fuller seems to have concluded from relativity that lowering the mass of buildings, vehicles, and other artifacts would increase the speed of production and free up energy for reinvestment in increased production. Even as he oriented his work toward this "time-rate" logic, though, Fuller invested the fourth dimension with the existential meanings and ethical imperatives that had accrued to the fourth dimension of space in the writings of Hinton, Ouspensky, Bragdon, and others. Fuller described the aim of his work as "the complete subjection of materialism to the will of the unselfish or spiritual man," and he claimed that the 4D house represented "harnessed—not worshiped materialism—true mind over matter—on the road from the complete, stony, compressive darkness of selfish materialism to the infinity of lightful, abstract, harmonic unselfishness."[20]

Technocracy

Fuller's 4D concept owed more to Henry Ford's system of mass production than it did to Einsteinian relativity. Fuller saw Ford as having translated Einstein's concept of a "time-rate world" into industrial practice to achieve great improvements in manufacturing efficiency. By standardizing the time in which workers performed automotive assembly operations, Ford was able to synchronize nearly the entire production process to run on a continuous assembly line. Fuller credited Ford with rationalizing automobile production based on "a TIMING system, a time-coordinated planning," to achieve unprecedented efficiency. The fourth dimension summed up the promise of industrialization withheld by such obstacles to progress as war, vested business interests, and government regulation. Referring to the Murphys, the Everyman family Fuller used as a rhetorical foil throughout *Nine Chains to the Moon*, he explained: "Einstein's 'relativity' reached the Murphys through Ford."[21]

In creating 4D housing, Fuller sought to apply Ford's timing system to the shelter industry. Given the central role played by time in nature and industry, he argued, "progressive design must be time saving."[22] By framing his work as the application of relativity theory to house design, Fuller characterized his "universal architecture" as an extrapolation of natural principles. In this he followed and deliberately echoed Taylor's discipline of scientific management. Taylor viewed himself as a scientist determining the principles of natural efficiency in order to apply them to industrial process. By identifying the "one best way" to solve any problem or perform any task, he argued, scientific management distilled, systematized, and replicated natural efficiencies.[23] Fuller adopted Taylor's philosophy, using the rhetoric of scientific management to naturalize his 4D architecture. "Nature," he explained in the *Time Lock*, "has in the course of time solved every mechanical problem" by segregating and solving functions. "Slowly nature has centralized production through industry, and taken the one best mechanical way of doing something . . . and made it available to all who will"—through devices ranging from razors and cars to hats, shoes, and stockings. "The home is the same," he concluded; "the house and all its functions are material and therefore solvable in but one best way."[24] The fourth dimension provided Fuller with a vocabulary for describing the application of scientific management time control to all of human society to realize the millennial dream of restoring humanity to an Edenic state.

Fuller based his ideas about time planning and mass production not only on the work of Taylor and Ford but also on the arguments of engineers in the Technocracy movement, a movement that sought to place engineers and other technical experts in charge of production and consumption decisions. Initiated during World War I by scientific management experts from Taylorist societies, and inspired by the writings of sociologist Thorstein Veblen and engineer Henry L. Gantt, the Technocracy movement was an outgrowth of Progressive Era social and political reform ideals.[25] Its advocates believed that new industrial methods had made possible levels of production sufficient to create an economy of universal abundance. This potential, however, was withheld by the selfishness of finance capitalists, who sought to maximize their own profits rather than to distribute the benefits of new technology as widely as possible, and by politicians, who served the interests of owners before those of consumers. This interpretation was amplified into a social program by a second generation that included Walter Rautenstrauch, a Columbia University engineering professor; Howard Scott, a technician; and Stuart Chase, a journalist. These technocrats saw engineers and other technical specialists as disinterested parties who, if given control of the production system, could reorganize and rationalize it for optimal production. By establishing a centralized command economy to coordinate production and consumption, these men aspired to eliminate the social unrest created by scarcity and resource competition. They envisioned a shift "from arbitrary power to scientific administration" that would yield "social harmony through . . . 'the organization of human affairs in harmony with natural laws.'"[26]

Scott was especially important to the elaboration of technocratic theory. As founder and chief engineer of a group called the Technical Alliance, which existed from 1919 to 1921, then again as a leading figure in the Committee on Technocracy, a research group established at Columbia University by Rautenstrauch in 1932, he declared that the potential for technologically produced abundance rendered obsolete all existing economic and political systems because they were based on scarcity. Scott conducted an energy survey of North America that analyzed the history of three thousand industries using a system of energy accounting that measured productivity and efficiency in terms of ergs, the basic unit of work, rather than in monetary terms. When the Committee on Technocracy split into two separate factions in 1933, Scott established Technocracy, Inc., through which he outlined proposals to transform the North American continent into an integrated command economy coordinated by an all-powerful hierarchy of experts called a "Technate."[27]

Fuller drew extensively from technocratic analyses in developing his own ideas and methods during the formative period of his career in the late 1920s and 1930s. In *Nine Chains to the Moon* he extrapolated his 4D theory of industrialism into a full-fledged technocratic theory of society and history. Fuller adopted most aspects of technocratic thinking, including the conviction that industrial potential was withheld by finance capitalists and politicians. Particularly influential on Fuller's thinking was Scott's theory that the efficiency with which a society converted energy from natural resources was the key index of human progress. The "Dymaxion Charts for Economic Navigation" appended to *Nine Chains to the Moon* were only the first of many projects in which Fuller adopted Scott's practice of industrial survey and energy accounting. They led to his eventual establishment of the World Resources Institute and sponsorship of the World Design Science Decade, 1965–75. With the infusion of technocratic theory and terminology, Fuller's 4D rhetoric acquired a greater degree of substance and scope, becoming a macrohistorical theory of society based on energy and industrialization.

Fuller encountered technocratic thinking through personal relationships with leading technocrats, including Scott, Chase, and the Committee on Technocracy member Frederick Ackerman, as well as with their less prominent associates such as the engineers Clarence Steinmetz and Irving Langmuir. He may also have read Veblen's articles in *The Dial*, the magazine cofounded by Emerson and Fuller's great aunt, Margaret Fuller Ossoli.[28] Fuller would later characterize himself as "a life long friend of Howard Scott and Stuart Chase" and explain that although never a member of Technocracy, Inc., he was "thoroughly familiar with its history and highly sympathetic with many of the views of its founders."[29] He drew on technocracy not only for its social theory but also for its model of professional organization. In 1932, as the Committee on Technocracy was garnering extensive press coverage and Scott was expounding its principles at Romany Marie's, the Greenwich Village restaurant at which both he and Fuller frequently dined together, Fuller founded a group of New York architects and housing reformers called Structural Study Associates, or SSA. Structural Study Associates was an architects' organization parallel to the Committee on Technocracy, as well as to the many other technocracy organizations founded in 1932 and 1933 throughout the United States and Canada. Devoted to increasing efficiency in the housing industry, SSA and its journal, *Shelter*, advocated technically sophisticated visions of industrial architecture for a high-technology society.[30]

Fuller's maturing vision of a 4D or Dymaxion society was in many respects an architect's reply to technocratic proposals for government by engineers. While Fuller shared the technocratic conviction that there existed only "one best way" to organize society for maximum production, he was too much of a libertarian to accept Scott's proposal that North America be ruled by a Technate empowered to determine everything from what would be produced to when a given employee would go to work. By the late 1930s, Fuller had repudiated Technocracy, Inc., calling it an "autocracy of engineers" and arguing that it was doomed because it allowed no room for individual speculation and initiative.[31] "There is a 'best' geometrical pattern and order constantly evolved in a scientifically unified establishment," he asserted in 1932, "without in any way detracting from the chances of life or the individual development of man."[32] Fuller's proposals for industrial design sought to reconcile his faith in "one best way" with his libertarian ethos. He set out to create "a whole new world industry concerned only with man's unavoidable needs and implementation of his inherent freedoms."[33]

Marketing Rationalization

In lieu of a technocratic command economy, Fuller envisioned a market-based social transformation driven by consumer demand for more efficient products and technologies. He aspired to bring to market a wide range of industrial products that would objectify technocratic social reform through mechanical technology. Fuller believed that under the right circumstances consumers would voluntarily buy into the project of rationalizing the housing industry, just as they had bought into automobility once Ford had made car ownership affordable and easy. Fuller's shorthand for this principle was the slogan "new forms rather than reforms." Rather than try "to reform man," he later recalled, "what I would do was try to modify the environment in such a way as to get man moving in preferred directions."[34] Fuller used beauty to stimulate consumer desire for products that would rationalize energy use. "I never work with aesthetic considerations in mind," he once explained, "but I have a test: If something isn't beautiful when I get finished with it, it's no good."[35] By using geometry to endow his designs with aesthetic appeal, Fuller put the persuasive power of beauty at the service of his social principles and cosmological convictions.

Fuller was committed to the rhetorical power of geometric regularity in ways that could distort his pursuit of efficiency. His major claim for the efficiency of geodesic structures, for instance, was that they enclosed the greatest volume of space with the least amount of material. This arbitrary criterion helped to preserve the visual and formal integrity of his spheres and hemispheres, but it had little relation to the overall efficiency of a building. In some cases Fuller's commitment to this narrow conception of efficiency led him to espouse significant inefficiencies of other kinds. A 1958 design-build studio at the University of Natal in Durban, South Africa, for instance, demonstrated Fuller's conviction that industrialized aluminum dome housing was a suitable global solution to human shelter needs. Fuller led a group of students in building a dome-shaped shelter out of aluminum sheets to replace the indigenous woven-grass *indlu* as a housing type for South Africa's Zulu population. Given the low cost of South African labor relative to industrial products, however, the aluminum shelter made little economic sense. Not only did it cost substantially more than its indigenous counterpart; it also performed poorly as an environmental control valve, averaging significantly higher interior temperatures than the breathable *indlu* in Natal's hot, dry climate.[36] A month-long architecture studio at North Carolina State College in 1952, where Fuller had led students in the design of an automated cotton mill, had yielded a similar outcome. Because the resulting project contained the entire production process within a geodesic dome, automated production lines had to snake up through a series of several floors of varying size, then back down to the trucks that distributed the baled cotton. The "90% Automatic Cotton Mill" exemplifies the way Fuller's commitment to structural efficiency and formal rationality could trump his commitment to economic rationality and production efficiencies. In studio after studio during these decades, Fuller led students in studying a problem and developing a solution—which invariably turned out to be a geodesic dome.[37] These pedagogical encounters tested the range of constructional possibilities offered by the geometries and structural principles Fuller was investigating more than they did the ways to optimize the benefits of industrialization for a worldwide human population.

Fuller's discovery of the rhetorical power of geometry dated from 1928, when he redesigned the 4D house from a rectilinear shape that echoed familiar housing types to its novel hexagonal form. Fuller did not publish these early designs with his *Time Lock* text, as he wanted to retain proprietary control over his innovations while he pursued patent registration. But he soon began to capitalize on the formal distinctiveness of the house, building models and producing drawings for exhibition that highlighted the design of the house more than its relation to his larger vision of a global "shelter service." After a sketch of the "Hexagonal House" was displayed at a Chicago restaurant and published in the *Chicago Evening Post*, Fuller arranged for a model to be exhibited at the flagship Marshall Field's department store. Rebranded as the "Dymaxion House," Fuller's design became a sensational attraction, in the tradition of other art and museum displays that helped to draw shoppers to department stores. A year after its development, the 4D house had become a marketing device for increasing sales of hats, shoes, and stockings—the standardized industrial products that had helped to inspire its creation. From its exhibition at Marshall Field's, the Dymaxion house moved into other media and display contexts. Shortly after its department store debut, the house was exhibited at the Harvard Society for Contemporary Art, then at the Architectural League of New York and other venues.

Fuller's redesign of the 4D house from a rectangular to a triangulated hexagonal plan had structural advantages. It standardized the radial members branching off the central mast while also exploiting the stability of the triangle. Fuller also claimed that users would benefit from his hexagonal plan, suggesting that this radial layout facilitated rapid movement through the house.[38] One diagram of the 4D house even analyzed its plan according to a geometry of time. Beyond these structural and layout efficiencies, however, the geometry of the 4D house distinguished it from competing proposals for prefabricated housing by giving it the aesthetic appeal that permitted its exhibition in art galleries and architecture magazines.[39] The triangular module carried over from the plan and cable stays to the pattern of window mullions on the facade, where it lacked either structural or temporal rationale. In one of Fuller's widely reproduced drawings, this triangular module carried over even into the layout of the caption, where it served not efficiency but aesthetics (fig. 7.2). Fuller used geometry to achieve not only structural and planning efficiency but also formal consistency and aesthetic power. Through its use in virtually all of Fuller's designs, the triangle came to symbolize the liquidation of conventional buildings, vehicles, maps, and cities by his industrially optimized, globally integrated 4D solutions.

Figure 7.2 Elevation, isometric, and plan of the single-family 4D house after its redesign based on a triangular module. Source: J. Baldwin, *BuckyWorks: Buckminster Fuller's Ideas for Today* (New York: John Wiley, 1996).

Sumptuary Design

Ford is famous for having said that the customer could have any color Model T as long as it was black. Fuller took a similarly restrictive approach to variety in housing design. Within Fuller's four-dimensional political economy, the greater human freedom permitted by technological optimization more than offset the limits to individual choice entailed in the standardization of housing. Freed from the necessity of working by increasingly efficient industry, and freed from the burdens of housekeeping by the rationalized house, women and men would find themselves with copious time for creative pursuits. "As we save time, and conserve it by shorter and more lasting methods and materials, we make time available to all the world, in the form of light, music, leisure for philosophic enjoyment; or . . . for the housing of our ever developing, finer, more creative selves."[40] The 4D house, Fuller predicted, would become "a place in which to live free from worry, free to explore, free to devise, include, refine, free to compose and synchronize."[41]

Fuller anticipated that his house design would change its occupants' purchasing patterns in other economic sectors, too. He incorporated into the *Time Lock* a long letter to George N. Buffington, a banker with whom he frequently discussed his ideas, recounting his visit to the opening of a new Woolworth store in Chicago.[42] The eagerness with which customers purchased frivolous items at the five-and-dime, Fuller explained, was "a result of a starved and 'don'ted' childhood in too close living quarters, with no other outlet for activity, amongst people whose minds have been purged of creative thought, as children." Fuller's 4D housing would eliminate this practice of consuming as a sublimated expression of creative energies stunted by childhood deprivation. "All the junk and temptation with which our store windows are crammed, furniture dealers, picture dealers, bird cage dealers, etc., etc.," Fuller asserted, "will vanish with 4D housing established. Purchase of this 'riffraff' will then be as out of the question as purchase of Morris chairs for Fords."[43] Instead of these "time wasting" objects, people would purchase the instruments of creativity. "Photographic supplies, sports equipment, tools, laboratory equipment, musical instruments, art supplies and any adjunct of creative or rhythmical activity will ever increase in sale." Fuller's industrial shelter service was a project of sumptuary regulation designed to change behavior not only in housekeeping and child rearing but also in home decorating, furnishing, and consumption more generally. The primary significance of the triangular module pervading the 4D house may well have been that it left no room for occupants to add home furnishings and decorative accessories expressing their personal taste. Fuller's design redefined personalized decoration as wasteful consumption.

The potentially intrusive nature of Fuller's sumptuary vision emerges in an early draft of the *Time Lock* as "Fuller's Law of Economics." This section, cut from the manuscript before it was mimeographed and sent out to potential sponsors and investors, suggests that individuals should receive monetary credit to the extent that their work saves time for others, but, conversely, they should be charged for any indulgence of personal appetite lacking socially redeeming purpose. Fuller called such self-indulgence "bestial" and associated it with the stomach. He envisioned his 4D world operating according to a "specific economy" of moral behavior that punished selfishness but rewarded the practice of working toward "acquisition of harmony." Through a complex monetary reward structure, he hoped to incentivize people to optimize the social value of their time usage. "When we have learned complete mastery of our selves to the extent of complete unselfconsciousness and unprocrastination in fulfillment of true duties as revealed," Fuller explained, "then we can control all other matter." It was just a matter of "licking the bestial self."[44]

Projective Ornament

The role that the beauty of regular geometry played in Fuller's regulation of consumption through design reflects his adaptation of strategies that Bragdon had developed in creating his system of projective ornament. An architect and critic with a substantial practice in Rochester, New York, Bragdon created projective ornament to serve as a universal ornamental language suitable to the full range of modern programs and contexts.[45] Troubled by the class antagonisms of industrial society, Bragdon criticized liberal modernity for being excessively individualistic and materialistic. His 1918 book *Architecture and Democracy*, for instance, condemned the new feudalism of industrial society for a pervasive lack of unity and beauty in American cities. In response, he enlisted architecture in the construction of a common culture. By integrating a society divided by distinctions of class, language, and national origin, projective ornament would turn architecture from a technique of differentiation and distinction into one of integration.

Bragdon turned to mathematics, and its formal expression through geometry, as the basis for a universal, orderly, and impersonal design language. Taking the fourth dimension as a newly discovered key to correct understanding of nature, he argued that four-dimensional geometry should replace both ornament based on the historical architectural styles and the novel ornamental motifs based on nature that were being used by midwestern progressives. To generate ornament from four-dimensional mathematics, Bragdon adapted the traditions that mathematicians had developed for representing elusive four-dimensional shapes such as the tesseract, the four-dimensional extrapolation of the cube. He unfolded cubes, tetrahedrons, icosahedrons, and other Platonic solids to generate two-dimensional figures, and he employed a variation on this technique to "unfold" tesseracts and other four-dimensional "hypersolids" into three dimensions. In similar fashion Bragdon used axonometric projection to graphically represent three-dimensional solids graphically, then adapted the technique to project four-dimensional shapes "down" two dimensions to generate additional graphic patterns (fig. 7.3).[46] By selectively accentuating and repeating elements of these different patterns and projections, Bragdon turned them into ornament (fig. 7.4).

Bragdon hoped that projective ornament patterns would teach viewers to see space as a Riemannian *n*-dimensional manifold. Projective ornament translated Bragdon's hyperspace philosophy into architecture and design by demonstrating the relation between the "lower spaces" of two and three dimensions and the "higher space" of four-dimensional communion. Bragdon associated the fourth dimension with Charles Howard Hinton's principle of "casting out the self," and the crystalline geometries of projective ornament gave formal expression to such "four-dimensional" social values as universality, impersonality, objectivity, and order. By disciplining sinuous arabesques to the strict lines of

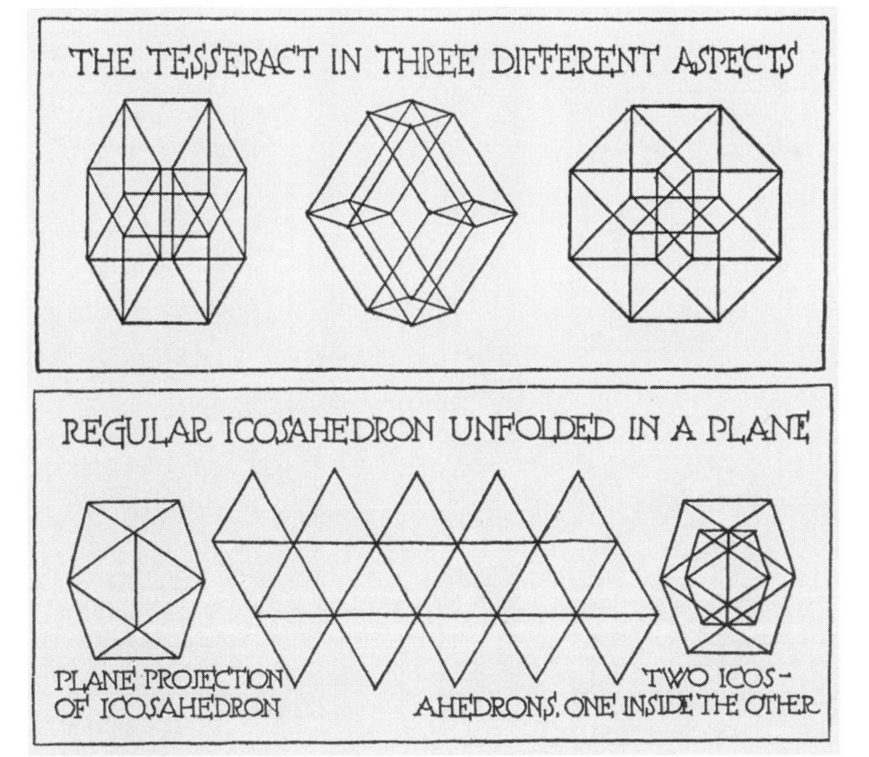

Figure 7.3 Diagrams demonstrating how three- and four-dimensional shapes could be represented on the two-dimensional page using unfolding and isometric projection. Source: Claude Bragdon, *Projective Ornament* (Rochester: Manas Press, 1915), 29, 43.

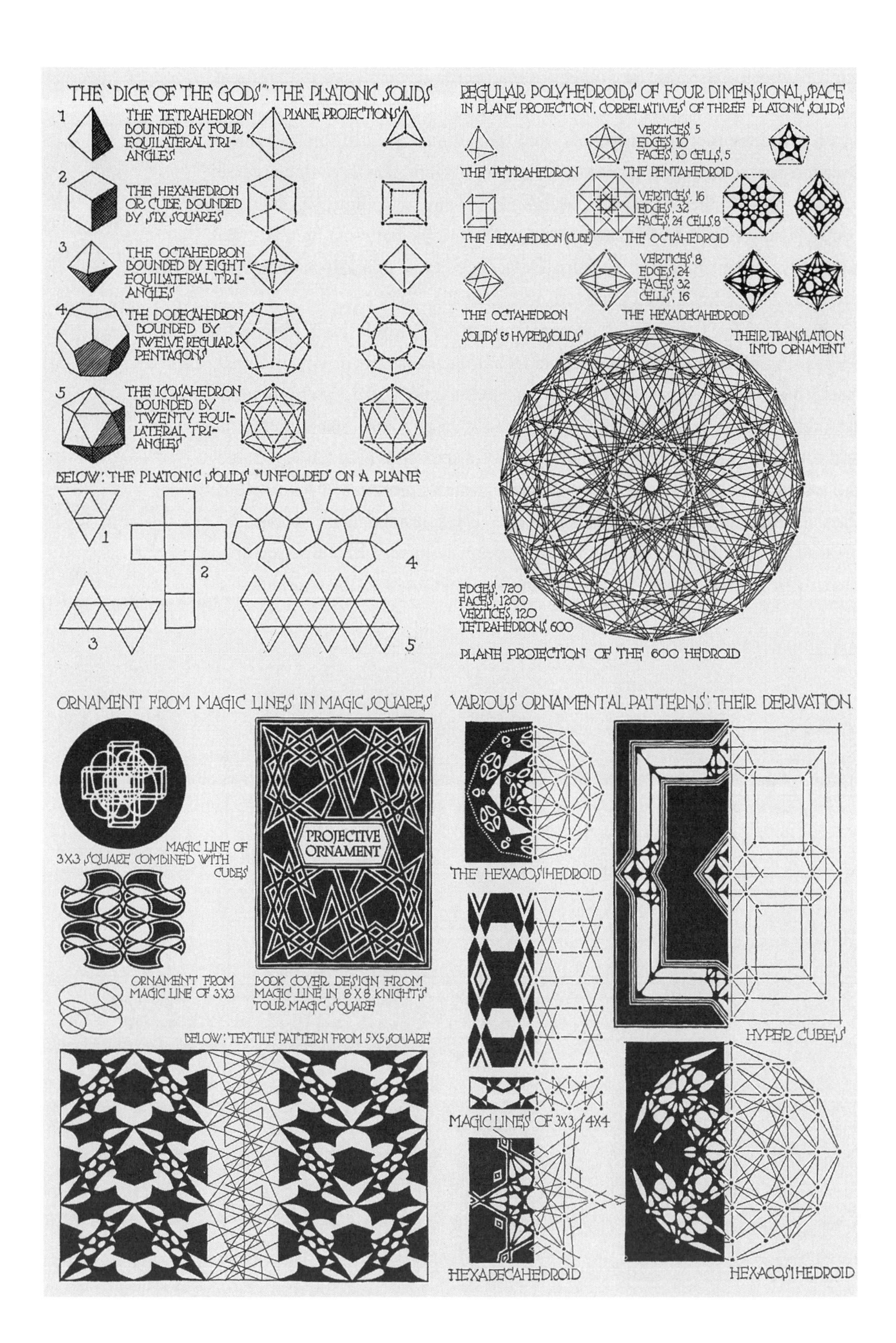

THE "DICE OF THE GODS": THE PLATONIC SOLIDS
1 THE TETRAHEDRON BOUNDED BY FOUR EQUILATERAL TRIANGLES
PLANE PROJECTIONS
2 THE HEXAHEDRON OR CUBE, BOUNDED BY SIX SQUARES
3 THE OCTAHEDRON BOUNDED BY EIGHT EQUILATERAL TRIANGLES
4 THE DODECAHEDRON BOUNDED BY TWELVE REGULAR PENTAGONS
5 THE ICOSAHEDRON BOUNDED BY TWENTY EQUILATERAL TRIANGLES
BELOW: THE PLATONIC SOLIDS "UNFOLDED" ON A PLANE
1
2
3
4
5
REGULAR POLYHEDROIDS OF FOUR DIMENSIONAL SPACE IN PLANE PROJECTION, CORRELATIVES OF THREE PLATONIC SOLIDS
VERTICES, 5 EDGES, 10 FACES, 10 CELLS, 5
THE TETRAHEDRON
THE PENTAHEDROID
VERTICES, 16 EDGES, 32 FACES, 24 CELLS, 8
THE HEXAHEDRON (CUBE)
THE OCTAHEDROID
VERTICES, 8 EDGES, 24 FACES, 32 CELLS, 16
THE OCTAHEDRON
THE HEXADECAHEDROID
SOLIDS & HYPERSOLIDS
THEIR TRANSLATION INTO ORNAMENT
EDGES, 720 FACES, 1200 VERTICES, 120 TETRAHEDRONS, 600
PLANE PROJECTION OF THE 600 HEDROID
ORNAMENT FROM MAGIC LINES IN MAGIC SQUARES
MAGIC LINE OF 3X3 SQUARE COMBINED WITH CUBES
PROJECTIVE ORNAMENT
ORNAMENT FROM MAGIC LINE OF 3X3
BOOK COVER DESIGN FROM MAGIC LINE IN 8 X 8 KNIGHT'S TOUR MAGIC SQUARE
BELOW: TEXTILE PATTERN FROM 5X5 SQUARE
VARIOUS ORNAMENTAL PATTERNS: THEIR DERIVATION
THE HEXACOSIHEDROID
HYPER CUBES
MAGIC LINES OF 3X3 / 4X4
HEXADECAHEDROID
HEXACOSIHEDROID

geometric crystals, Bragdon allegorized the individual's willing submission to the demands of a higher necessity. Embracing the association that the English critic John Ruskin had made between regular geometry and servility, Bragdon used ornament to promote a sumptuary ethos. By emphasizing the capacity of crystalline geometry and repetitive all-over patterns to create a consistent and orderly aesthetic totality, Bragdon advocated what he called "exquisite acquiescence": the beauty of individual submission to the demands of "beautiful necessity," an Emerson phrase that Bragdon adopted as a mantra.[47] Because its patterns could be equally well realized in two or three dimensions, projective ornament also formed a bridge between architecture and design in other media, including typography, advertising, textiles, painting, and film. Bragdon's illustrations of how his ornament could be applied to design, meanwhile, emphasized its capacity for creating aesthetically unified environments (fig. 7.5).

From Projective Ornament to Comprehensive Design

Bragdon's writings were a major source of Fuller's hybrid fourth dimension. Fuller probably discovered Bragdon through *The Dial*, which from 1917 into the 1920s published his essays, reviews, and projective ornament designs.[48] When he mailed out copies of the *Time Lock* manuscript in 1928 to key figures in industry, education, architecture, literature, and criticism, Fuller sent one to Bragdon, along with a letter explaining that Bragdon's "books . . . , researches and associations with the matter" addressed by the manuscript made his "study and comment on it" imperative.[49] Fuller's letter specifically mentioned *Architecture and Democracy*, along with Bragdon's translation of Ouspensky's *Tertium Organum*. *Architecture and Democracy* was also the third item on the handwritten "List of References" that formed an appendix to the *Time Lock*.

Fuller adapted the ethics of the fourth dimension of space as developed by Hinton, Bragdon, and Ouspensky to a relativistic understanding of the fourth dimension as time. His vision of social harmony through time controls updated Bragdon's pursuit of social harmony through space controls, much as his characterization of art as "an harmonic division, composition, and projection of *time*" reformulated Bragdon's characterization of architecture and design as "rhythmic space subdivision."[50] Traces of hyperspace philosophy occur throughout Fuller's work. One is Fuller's description, in *Nine Chains to the Moon*, of the mind as a "phantom captain" that exists independent of the body (its ship) and possesses an intuitive "awareness of perfection" against which to measure phenomena. This closely followed Hinton's and Bragdon's characterizations of consciousness as a four-dimensional captain piloting the "battleship" of the body.[51] Another is Fuller's use of Emerson to describe the rhetorical power of

Figure 7.4 (opposite) Illustrations summarizing Bragdon's system of projective ornament by showing how projections and unfoldings could be turned into ornamental designs. Source: Claude Bragdon, *The Frozen Fountain*, © 1932 and 1960 by Henry Bragdon. Used by permission of Alfred A. Knopf, a division of Random House, Inc.

"necessary" beauty, which echoed Bragdon's Emersonian rhetoric of "beautiful necessity." Fuller adapted Bragdon's disciplining of arabesque to crystal in an illustration from *Nine Chains to the Moon* that shows the progressive straightening of a bent line, representing the individual motivated negatively by fear, into a tautly "tensed" line, representing the individual motivated instead by positive longing. Fuller even referred obliquely to Bragdon's system of projective ornament in the *Time Lock* when he observed: "It is a strange paradox that the world turns up its nose at the artist's projection of natural or fourth dimensional matter in three dimension or cubistic form, and yet goes on designing its houses about itself in this same limited cubism."[52]

Given that Bragdon's writings were a significant source of Fuller's understanding of the fourth dimension, it would be surprising if his system of projective ornament played no role in shaping Fuller's thinking about the socializing power of geometry and beauty. Fuller suggested a number of sources for his mast-hung, tension-cabled hexagonal aluminum house, including lighthouses, ship masts and conning towers, pagodas, octagon houses, and suspension bridges. The Swiss architect Le Corbusier's modernist houses, published in his book *Towards a New Architecture*, probably suggested additional features of the 4D house, such as the carport, roof deck, and horizontal window bands. The geometries Fuller employed in the house and in subsequent projects, as well as the techniques he used to manipulate those geometries, however, duplicated those that Bragdon had used to represent nature in generating projective ornament patterns (figs. 7.6, 7.7). Unfolded

Figure 7.5 (opposite) Illustration showing how projective ornament can integrate buildings with furnishings, textiles, and even clothing.
Source: Claude Bragdon, *Projective Ornament* (Rochester: Manas Press, 1915), 14.

Figure 7.6 (below) Self-portrait showing Bragdon using wire models to study the relation between three- and four-dimensional shapes.
Source: Claude Bragdon, *The Frozen Fountain*, © 1932 and 1960 by Henry Bragdon. Used by permission of Alfred A. Knopf, a division of Random House, Inc., 106.

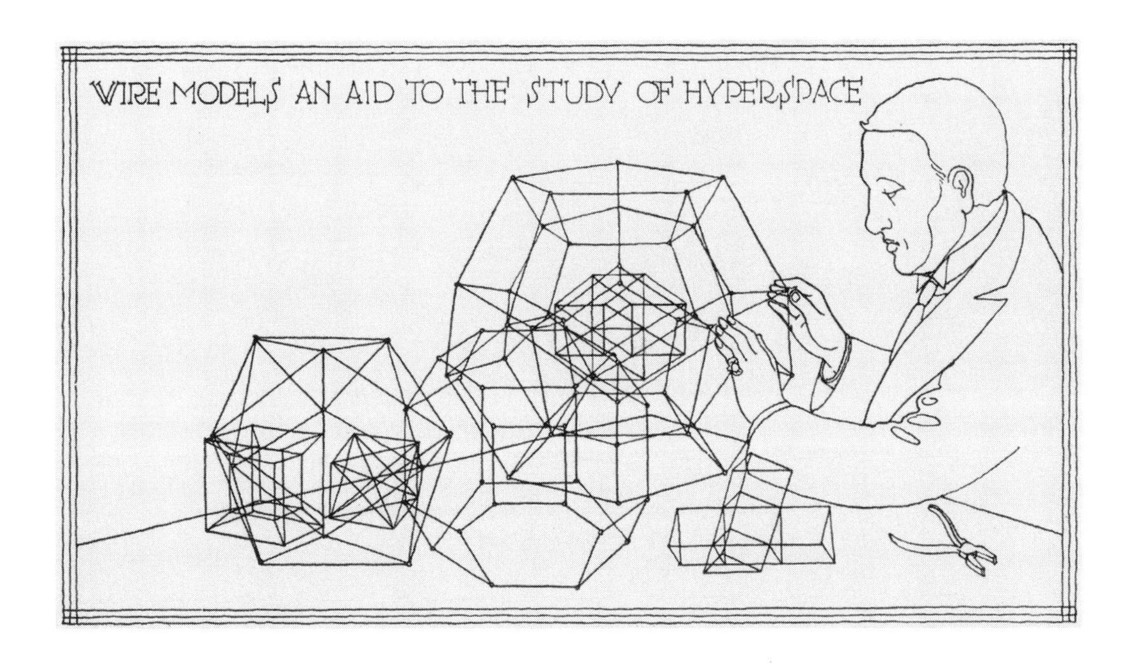

Figure 7.7
In developing the principles of geodesic structure at Black Mountain College in 1949, Fuller experimented with models made of cardboard, Venetian blinds, and other materials.
Source: Special Collections, Stanford University Libraries.

and projected solids and hypersolids in Bragdon's books, such as a projection of the four-dimensional "hexadekahedroid" in *Architecture and Democracy* or the triangular and tetrahedral patterns throughout Bragdon's work, are likely sources of Fuller's shift from rectangular to hexagonal planning in the 4D house. Bragdon's many diagrams showing how to translate three-dimensional volumes into two-dimensional images reappeared in Fuller's system of Dymaxion projection and in the geoscopes that Fuller constructed with John McHale during the 1950s and 1960s (fig. 7.8). Bragdon established the association between geometric projection and four-dimensional discipline that informed much of Fuller's work.

Bragdon continued during the late 1920s and early 1930s to explore the symbolic meaning of geometry in ways that anticipated Fuller's later designs. His 1932 treatise *The Frozen Fountain* featured polyhedra that would become central to Fuller's later work, including the soccer-ball shape that would come to be called a "buckyball" because of its geodesic properties (fig. 7.9). The *Frozen Fountain* endpapers were even decorated with a drawing showing the

Figure 7.8
Drawing of the icosahedral geoscope built in 1964 by students of John McHale at the University of Colorado, Boulder, based on the Dymaxion air-ocean map projection, in which the globe is mapped onto an icosahedron and then unfolded.
Source: John McHale, *World Design Science Decade 1965–1975 Phase I (1965) Document 4: The Ten Year Program* (Carbondale: World Resources Inventory, Southern Illinois University, 1965).

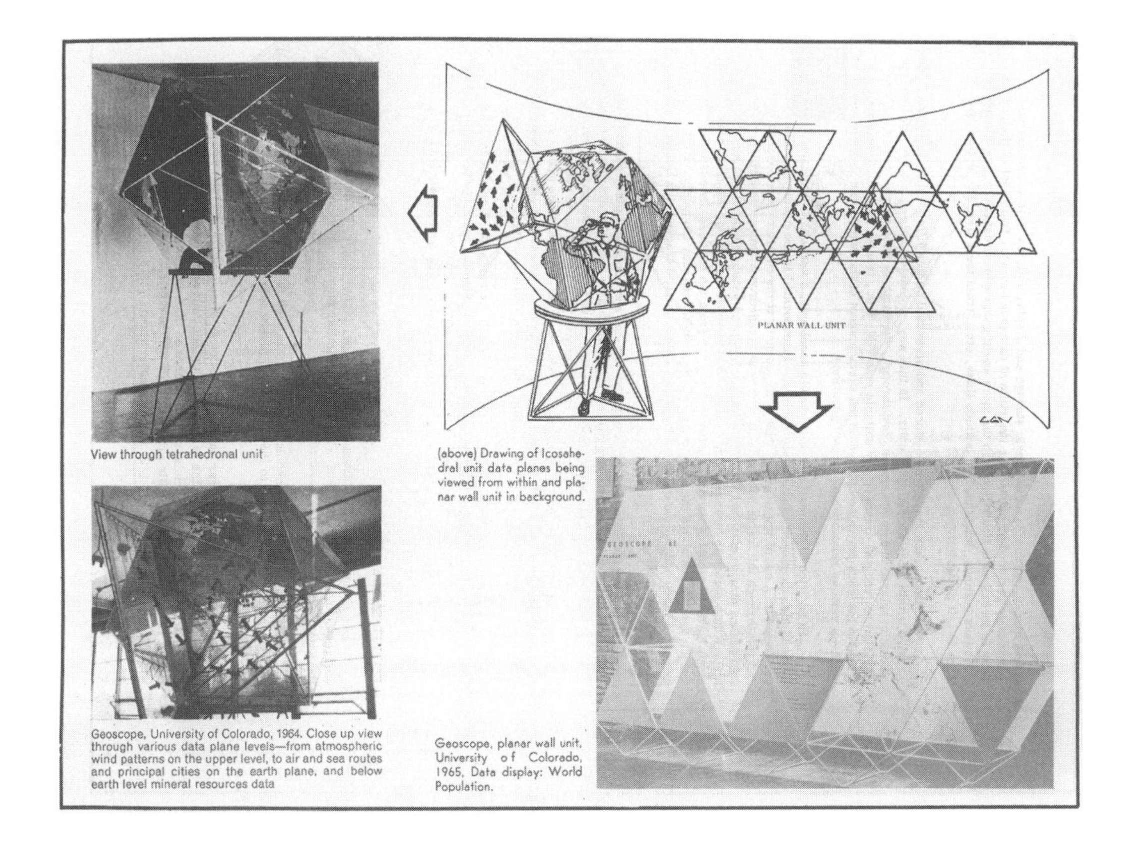

allegorical figure of Sinbad, who represented the creative artist, taking shelter in an icosahedral cage constructed of struts and ball connectors—yet another anticipation of geodesic structures (fig. 7.10). Fuller's evolving exploration of the structural, cosmological, and aesthetic potential of geometry may have been informed by these drawings or by others that Bragdon exhibited in a 1941 New York gallery show under the title "Mathematical Abstractions." Fuller even decorated the dome home he and Anne shared in Carbondale, Illinois, with what can only be called projective ornament (fig. 7.11).

Figure 7.9 Bragdon's design for an entryway in projective ornament, published in *The Frozen Fountain*, featured icosahedral and polyhedral lanterns in the soccer-ball form that would come to be known as the "buckyball," along with patterned glass, curtains, ceiling, and floor.
Source: Claude Bragdon, *The Frozen Fountain*, © 1932 and 1960 by Henry Bragdon. Used by permission of Alfred A. Knopf, a division of Random House, Inc.

Figure 7.10
The endpaper of *The Frozen Fountain* featured Sinbad, an allegorical figure representing the architect or artist, taking shelter in an icosahedral ball-and-strut shark cage.
Source: Claude Bragdon, *The Frozen Fountain*, © 1932 and 1960 by Henry Bragdon. Used by permission of Alfred A. Knopf, a division of Random House, Inc.

Despite the great differences in scope between Bragdon's projective ornament and Fuller's designs for houses and geodesic structures, there are substantial continuities between these two bodies of work, both of which used geometry to translate a four-dimensional sumptuary ethos into design. Fuller's commitment to geometry for a combination of structural, planning, and aesthetic reasons was based on its potential for effectiveness in rationalizing production and consumption through the liberal mechanisms of

Figure 7.11 Fuller hosting students inside his Carbondale dome home in 1961, with a polyhedron projection ornamenting the ceiling.
Source: Special Collections, Stanford University Libraries.

democratic polity and a market economy. From 1928 his work combined a technocratic commitment to efficiency and planning with an architectural tradition of sumptuary regulation through ornamental pattern. By synthesizing these strategies through geometry in the 4D house, Fuller aimed to show how the industrialization of housing could discipline selfish consumption and so "validate the libertarian principle without recourse to the temporary efficiencies of dictatorship."[53]

8

Backyard Landing

Three Structures by Buckminster Fuller

Maria Gough

"Let me first congratulate you on the perfectly splendid showing at the Museum of Modern Art," enthused *Time* magazine's architectural critic Cranston Jones in a letter to R. Buckminster Fuller of September 23, 1959. "Space structures for the twentieth century never looked better than in the sculpture garden."[1] The critic was referring to *Three Structures by Buckminster Fuller* (fig. 8.1), which had opened the day before and would run for more than a year without anybody quite intending, finally closing in November 1960. Installed on what was then an empty lot adjacent to the museum's sculpture garden, the exhibition comprised demonstration models of three spectacular structures: the *Octet Truss*, *Tensegrity Mast*, and *Geodesic Rigid Radome*. Common to all three was a stability derived from geometry rather than mass; this geometry was based on triangulation rather than on the square or the cube, the Bauhausian building blocks that Fuller reputedly despised.

While *Three Structures* was not Fuller's first engagement with MoMA's empty lot—his unusual *Dymaxion Deployment Unit* had appeared there in 1939[2]—it was nevertheless the museum's largest ever commitment of space and time to the exposition of his work. This commitment came at an important transitional moment in Fuller's career. With considerable commercial success to his name, due to military and government contracts in the 1950s, Fuller was now searching for potential applications for the three exhibited structures. It was also an important moment in the life of the museum's Department of Architecture and Design, the critical orientation of which was undergoing a pluralization if not a sea change with a newly appointed young director, Arthur Drexler, now at the helm. Having been a major institutional propagandist for modernist architecture ever since its first director, Philip Johnson, collaborated with Henry Russell Hitchcock on the groundbreaking *International Exhibition of Modern Architecture* in 1932, the MoMA department was now beginning to move in new directions.

Despite the historical significance of *Three Structures* for the designer and the museum alike, the exhibition has never been revisited. Drawing predominantly

Figure 8.1 View of *Three Structures by Buckminster Fuller*, Museum of Modern Art, New York, 1959–60.
Source: Special Collections, Stanford University Libraries.

on archival materials, I examine MoMA's mediation of public understanding of Fuller's design science with respect to two major problems. First, I address the ways in which *Three Structures* reconceptualized function, a crucial issue in the study and practice of modern architecture and design, within a new discourse of futurity. Second, I take up the complex problem of authorship in Fuller's enterprise by foregrounding the existence and significance of a much less well known exhibition, *Buckminster Fuller: Supplementary Exhibition of Models, Photographs, and Drawings.* Also held at MoMA in 1959, this companion exhibition thickened the plot of *Three Structures* by including the work of Fuller's collaborators and former students.

I

For a couple of decades at midcentury, the Museum of Modern Art ran an innovative exhibition program right in its own backyard, inviting major architects and designers to build full-scale demonstration models of current projects. The hope was that by affording visitors direct, phenomenological experience of built forms, the museum would facilitate greater popular understanding of architectural space than that provided by photographs, drawings, and scale models alone.[3] One of the first projects, an elegantly functional modernist house for a suburban middle-class family, was designed in 1949 by Marcel Breuer, an expatriate Hungarian architect with the impeccable Bauhaus credentials most prized by the department's then directors, Johnson and Peter Blake. Functionality and the potential for immediate application were major curatorial concerns: "Had MoMA commissioned . . . [Buckminster] Fuller . . . to design such a house," Blake later mused, "critics would have accused it of being . . . out of touch with reality."[4] Thriving in the interstices of architecture, engineering, geometry, philosophy, and neologism, Fuller was too far out for such a mainstream brief. Yet, just ten years after Breuer's house in the garden, Fuller was busy on the very same midtown site.

If Johnson and Blake had foregrounded the functionality and feasibility of Breuer's demonstration model, Drexler considered the function of Fuller's structures in a much more open-ended, speculative fashion. According to a press release issued a month before the opening, the purpose of *Three Structures* was "to illustrate the extraordinary strength and lightness of Mr. Fuller's method of construction which utilizes the forces of tension and compression in an unconventional way and which may in time change the appearance and character of our buildings."[5] As such, *Three Structures* showcased neither specific functions nor even buildings but rather new structural systems and construction methods that seemed to promise the future of architecture. Drexler's introductory label informed exhibition visitors that Fuller "believes that the designer's real responsibility no longer is the creation of individual buildings or objects, but rather . . . the interrelating of physics, mathematics, and [humanity's] well-being."[6]

Designed especially for the exhibition, Fuller's super-scaled one-hundred-foot-long, thirty-five-foot-wide, and twenty-four-foot-high space frame dominated the MoMA site (fig. 8.2). The arrangement of its gold anodized aluminum tubes into octahedrons (eight-sided figures) and tetrahedrons (four-sided figures) lent the structure its invented name, *Octet Truss*. From a single off-center support, the truss cantilevered sixty feet in one direction and forty in the other. Drexler alerted visitors to the fact that the *Octet Truss* was "not an actual 'building'" and thus had no specific function as such. Nevertheless, he continued, "its structural principle can be used wherever it is necessary to make large uninterrupted roofspans: concert halls, factories, museums, train sheds,

Figure 8.2 R. Buckminster Fuller, *Octet Truss*, Museum of Modern Art, New York, 1959–60. Photo by Alexandre Georges. Source: Special Collections, Stanford University Libraries.

airplane hangars." Furthermore, "the nature of [its] structural system suggests that we may ultimately learn how to 'weave' enormous buildings that will differ in every way from what we now call architecture."

Also designed especially for the exhibition was the *Tensegrity Mast* (fig. 8.3), a thirty-six-foot tower of aluminum tubes and thin monel rods that demonstrated a novel structural system of discontinuous compression and continuous tension. Whereas in the truss the aluminum tubes handled both tensile and compressive forces, in the mast these same tubes carried exclusively those of compression, with tensile forces residing in its remaining members. The compression members thus became small islands in a sea of tension, as Fuller liked to put it. (The term *tensegrity* was a neologism he formed by contracting his original term for the system, "tension-integrity.") The startling achievement of the discontinuous compression–continuous tension system was its suspension of rigid elements in space by means of tension alone. "Although each unit appears to be carrying the one above it," Drexler wrote, "they are more accurately described as holding each other apart."[7] (Popular reports repeatedly described the mast as an engineer's version of the Indian rope trick.) While the *Tensegrity Mast* had "no practical application," it was, Drexler asserted, "theoretically possible to use this system in construction building of enormous scale."[8]

The third structure was a fifty-five-foot diameter greenish-yellow translucent plastic and fiberglass *Geodesic Rigid Radome* (fig. 8.4), a special-case iteration of the dome structure with which Fuller was to become ubiquitously associated. The radome was assembled by bolting adjoining panels together so that skeleton and skin were one. Of the three structures on display, the *Geodesic*

Figure 8.3
R. Buckminster Fuller, *Tensegrity Mast*, Museum of Modern Art, New York, 1959–60.
Photo by Alexandre Georges. Source: Special Collections, Stanford University Libraries.

Figure 8.4
R. Buckminster Fuller, *Geodesic Rigid Radome*, Museum of Modern Art, New York, 1959–60.
Photo by Alexandre Georges. Source: Special Collections, Stanford University Libraries.

Rigid Radome was the most controversial with respect to function. Drexler's wall text noted that "this particular dome is used to house radar installations on the Artic Distant Early Warning Line," where it withstands Artic winds of up to two hundred miles an hour despite weighing only twelve hundred pounds.[9] The DEW Line was a vast radar surveillance system built in the 1950s and stretching three thousand miles from the northwest coast of Alaska to the eastern shore of Baffin Island opposite Greenland, whose purpose was to warn Canada and the United States of impending air strikes from the Soviet Union. The great success of this and other military applications of the radome played a crucial role in the development of Fuller's design science because government royalties provided the polymath with substantial means to fund other, more speculative projects: According to Yunn Chii Wong, the radome was the most profitable of Fuller's dome enterprises, earning the designer an estimated two million dollars in royalties between 1954 and 1961.[10] But Drexler was not especially concerned with the radome's historical function as a defensive weapon of the cold war nor with how it had helped to advance Fuller's career. Instead, he emphasized the radome's potential future in the form of, for example, "T.V. studios, ball park coverings, swimming pools, houses, [and] bomb shelters,"[11] thereby realigning it with the as-yet-unrealized functional potential of the other two structures on display.

On the MoMA lot the radome came to fulfill a variety of more modest functions, such as an outdoor shelter for the museum's annual Children's Carnival, and as the location for a *Town & Country* magazine fashion shoot in which the New York philanthropist Mrs. Bertrand Taylor III sported a Handmacher suit, Lilly Daché hat, Crescendoe gloves, and a Lennox patent leather bag (fig. 8.5).[12] During the winter of 1959–60 the radome also served as an unheated studio workshop for the Swiss artist Jean Tinguely's preparation of *Homage to New York*, a one-time performance event that took place in MoMA's garden in March 1960 (fig. 8.6). Involving as it did Tinguely's laborious construction of an elaborate junk machine that then destroyed itself during the performance, this last usage was particularly choice: nothing could be more antagonistic to Fuller's optimistic technophilia than Tinguely's nihilistic *Homage*. At one point Drexler even proposed using the radome as a screening

MRS. BERTRAND TAYLOR III, left, of New York, poses for "T&C" inside the Geodesic Rigid Radome invented by Richard Buckminster Fuller and on display at the Museum of Modern Art. Her box-jacketed suit is a black-and-white check of Arnel blended with silk, cotton, and rayon—a Crown Fabric. By Handmacher, at Lord & Taylor; Halle's; Scruggs Vandervoort Barney. Daché hat; Lennox patent leather bag; Crescendoe gloves. MRS. BROOKE TAYLOR, right, studies a Calder mobile in the new Park Avenue branch of The Chase Manhattan Bank, designed by Skidmore, Owings & Merrill. Her suit of Forstmann flannel shows the ultimate length reached by the longer jacket—this one pearl-buttoned, patch-pocketed. A Robert Knox design for Ben Gershel at Saks Fifth Avenue; Neiman-Marcus. Toby Coppock hat; gloves by Superb

NEW MOODS IN MANHATTAN

Figure 8.5
"New Moods in Manhattan."
Source: *Town and Country Magazine*, March 1960.

Figure 8.6 Jean Tinguely, *Homage to New York*, Museum of Modern Art, 1960. Fuller's *Geodesic Rigid Radome* is visible in the right background. Source: Special Collections, Stanford University Libraries.

room for Charles and Ray Eames's film *Glimpses of the USA*, which had made a big splash in Moscow just the previous summer when it was projected in one of the U.S. pavilions, an aluminum Fuller dome. Such elasticity of function was embraced by Fuller for both philosophical and pragmatic reasons. But these sundry functions in MoMA's backyard were largely serendipitous and coincidental to Drexler's deeper curatorial purpose.

A sense of Drexler's purpose may be gleaned from the exhibition's catalog, *Three Structures by Buckminster Fuller in the Garden of the Museum of Modern Art, New York* (fig. 8.7), which consists of a single sheet measuring thirty-six inches by twenty-four inches. Folded in half lengthwise and then again in three, the sheet provides six twelve-inch square pages for the reproduction of installation shots by the architectural photographer Alexandre Georges, elevation drawings and photo collages by Fuller, and commentary by Drexler. In its very design, this slim catalog thus tropes the modularity, seriality, and less-is-more ingenuity of the structures it accompanied. What is most interesting about the catalog is not so much Drexler's commentary, which was recycled from his wall texts and other exhibition-related publicity, but his selection of installation photographs. Georges had taken both day and night shots of *Three Structures*, but it is his spectacular night photographs that have the lion's share of space in the catalog.

If one unfolds the catalog fully, the entire verso of the sheet consists of a single nighttime mise-en-scène, in which a suited man is staged within the space of the exhibition (fig. 8.8). The dramatic contrast of raking light and deep shadow at firsts suggests film noir. Unlike a film noir character, however, the solitary figure does not lurk in the shadows but stands directly in the path of the light flooding from the dome, which glows like a giant silkworm, rendering the true genre of this scene science fiction. Standing just outside the dome's entrance, the man contemplates the superskeleton cantilevering overhead, which casts a geometric maze of shadows on the apartment building across the road. Spotlights punctuate the darkness; trees are ablaze with tea-lights. Henri Matisse's bronze *Backs* retreat to dark slabs on the rear wall. A modest wooden fence and gate is all that separates this would-be film set from 54th Street, which fronts MoMA's backyard as it existed then. Georges' architecture of the night withdraws the radome from the cold war reality that had fostered its production in order, instead, to spectacularize *Three Structures* as science fiction.

Figure 8.7
Front cover of Arthur Drexler, *Three Structures by Buckminster Fuller* (New York: Museum of Modern Art, 1960).
Source: Special Collections, Stanford University Libraries.

Figure 8.8
Verso of unfolded *Three Structures by Buckminster Fuller* exhibition catalog (New York: Museum of Modern Art, 1960).
Photo by Alexandre Georges.

Staging the scene of the exhibition as sci-fi fantasy was not idle speculation on the photographer's part but rather a prescient interpretation of Fuller's then-contemporary concern to relaunch the geodesic dome, already a leading protagonist within the military and trade-fair circuit, within a new discourse of futurity. The designer's "future [dome] projects include sky islands for use as launching platforms for missiles, underwater islands as bases for submarines, oil drillers for oceanographic surveys," the museum's publicity informed the press. "Domes might also be used to enclose whole cities on earth or the moon," read the wall text.[13] Fuller's most fantastic projection along these lines was the *Manhattan Island Dome*, which he was working on in 1959, concurrently with the preparation of *Three Structures.* Fuller presented his idea of superimposing a two-mile-wide geodesic dome over part of Manhattan in a photo collage (fig. 8.9) that had a prominent place in the supplementary exhibition[14] and was also reproduced in the catalog for *Three Structures* with the following caption: "The plan of a dome superimposed on part of Manhattan Island illustrates a possible use for large-scale structures. By enclosing so great an area it would become possible to dispense with much weatherproofing and protection now necessary for individual buildings. A dome of this size might be constructed

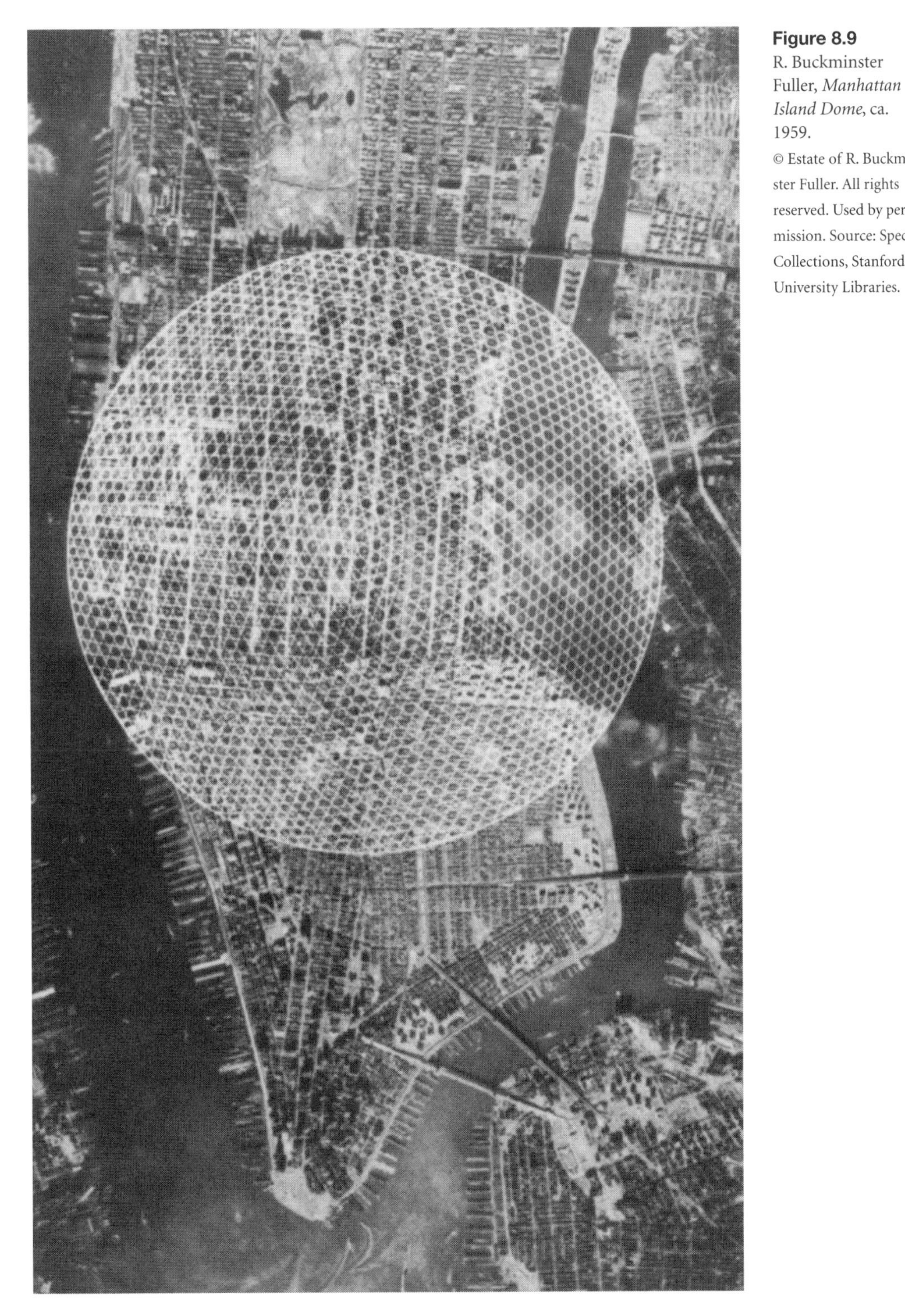

Figure 8.9
R. Buckminster Fuller, *Manhattan Island Dome*, ca. 1959.
 Source: Special Collections, Stanford University Libraries.

on the principle of the tensegrity mast, the tubes and rods being arranged in a configuration resembling that of the octet truss."[15] Glossed as such, the Manhattan Island project was presented to visitors as a fantastic summa of the three demonstration structures rolled into one.

"Looking far ahead, the museum visualizes a fantastic new world," reported Ada Louise Huxtable, the *New York Times* architectural critic, who quoted Drexler at length in her review of *Three Structures*: In Fuller's hands "buildings might no longer be a series of separate boxes with people moving from one to another," the curator had asserted. "The infinite clear spans suggest a new kind of shelter—vast domes enclosing entire communities, permitting continuous control of the environment. In effect, the city would be one building, with its necessary functions accommodated quite differently than they are today. We could climate control and reclaim whole areas of the Sahara, or of the Artic."[16] Over the next few years Fuller proposed a series of megastructures spinning off from the Manhattan Island dome, such as giant spheres one mile in diameter in which thousands of people would travel around the world, occasionally anchoring themselves atop mountains (*Cloud Nines* [1961]) and a giant floating tetrahedron that could be anchored anywhere its inhabitants chose (*Tetrahedronal City* [1965]).[17]

That the curatorial purpose of *Three Structures* was in part to present Fuller as a futurist rather than a functionalist is further supported by the fact that Drexler sought to include the designer in *Visionary Architecture*, an international group show on which he was working contemporaneously with his organization of *Three Structures*.[18] In a "long overdue" letter of February 1960 to Fuller's company, Synergetics, Inc., Drexler returned to the subject of the proposed *Visionary Architecture* exhibition, which he had first raised "many months ago." His ambition was "to call attention to a kind of underground history of architecture." To this end "projects [would] be selected on the basis of their being *unbuildable*," whether owing to technical difficulties or to the absence of a "social program to justify the architect's vision." Drexler argued that it was the "inability to define what would be socially desirable, let alone facilitate it"—and not technical limitations per se—that had prevented twentieth-century visionary projects from coming to fruition.[19]

Like his predecessors in the department, Drexler sought to encourage the museum visitor's phenomenological experience of architectural form but, in this case, of the unbuildable kind: "I would like to end the exhibition with a model so large that the public can walk through it. Such a model would occupy roughly a space 22 feet square. The floor or ground level of the model would be at chin height and the public would walk through a narrow trench. It might thus be possible to create the illusion of enormous distances and give to even the most visionary of projects a kind of Alice-in-Wonderland quality." Drexler welcomed suggestions from Synergetics as to how to accomplish this full-scale model of the unbuildable, providing their suggestions were "what the uninitiated would call crazy." He closed his letter with an apposite reference to *Three Structures* as "the *vision* in our garden."[20]

II

Buckminster Fuller: Supplementary Exhibition of Models, Photographs, and Drawings opened on October 28, 1959, in the MoMA's Far West Gallery and closed a little shy of a month later on November 23, 1959 (fig. 8.10). Also curated by Drexler, the exhibition was designed by the department's assistant director, Wilder Green. Fuller seems not to have been directly involved in its organization, at least not personally, since he requests information about it from the museum's publicity director: "I am eager to hear about and receive photographs and mimeographs of the room inside the Museum being

Figure 8.10
View of *Buckminster Fuller: Supplementary Exhibition of Models, Photographs, and Drawings*, October–November 1959. Photo by Kenneth Snelson. Used by permission. Source: Special Collections, Stanford University Libraries.

organized with models, etc., of the work of my associating companies and individuals," he writes on October 21, 1959.[21] No checklist of the exhibition appears to have survived, but an excellent sense of its contents may be derived from some ten installation photographs by Kenneth Snelson that are preserved in the museum's archives.[22]

The supplementary exhibition served to contextualize *Three Structures* in important ways. First, it situated Fuller's "vision in the garden" within the overall trajectory of his work. To shed light on the designer's inventions before the advent of the geodesic dome in the late 1940s, Drexler displayed the aluminum and plastic scale-model of the Fuller house (a.k.a. *Dymaxion Dwelling Machine, Wichita, Kansas* [1944–46]) from the museum's permanent collection, along with photographs of, for example, his three-wheeled Dymaxion car. A number of wood, plastic, and paper models pertaining to Fuller's explorations in the realm of geodesic geometry in the late 1940s and 1950s, like the *Wooden Hex Pent Trussed Geodesic Sphere*, were suspended from the ceiling or arrayed in vitrines. Documentary photographs showed earlier incarnations of the geodesic dome, such as the cardboard version designed for the Marine Corps, and of the *Octet Truss*, the structural system of which had first been used (though covered) in 1958 in the Union Tank Car Company dome in Baton Rouge. The *Octet Truss* was itself represented by a scale model, while one potential application of its principle was suggested by an intricate wooden model of a proposed *Airplane Hangar* (1955, visible in the left middle-ground of fig. 8.10). The photo collage for the *Manhattan Island Dome* project enjoyed its own wall.[23] Another section of the gallery documented the process of constructing the demonstration structures on display outside. Overall, the Far West Gallery gave the visitor a much broader understanding of Fuller's purpose in *Three Structures* than that afforded by the outdoor show alone.

Second, and controversially, the supplementary exhibition drew attention to the extent to which Fuller's process of invention was a collaborative one. As Alex Soojung-Kim Pang writes, "No inventor works without collaborators or in isolation from society." Despite the myth of the lone inventor on which Fuller's reputation thrived, his inventions were produced within a "network of consulting firms, industries, and universities that provided him with projects, resources, and labor."[24] In the organization of *Three Structures* Drexler was fully aware of this network. For example, in a letter to J. A. Vitale regarding the loan of the radome to the museum, Drexler confirmed that "signs in the exhibition and attendant publicity releases will bear the information that the radome structures were developed and tested by MIT Lincoln Laboratory for the U.S. Air Force."[25] True to his word, Drexler saw that all the exhibition publicity for

Three Structures mentioned the constitutive role of Lincoln Labs in moving Fuller's design off the drawing board and into production. Similarly, Drexler acknowledged that the Canadian company Aluminum Limited, who had acted as consulting engineers for the *Octet Truss*, had "contributed" or "co-sponsored" its design, while the *Tensegrity Mast*, he informed visitors, was "built" by Shoji Sadao and (the lighting designers) Edison Price, Inc.

But the matter of collaboration in *Three Structures* was more complex than merely one of fabrication. By the time of the exhibition, the proper name *Fuller* referred as much to a corporation as to an individual. Headquartering his operations in Long Island, Fuller had opened franchises ("Fuller Research Foundations") in Chicago, Detroit, Montreal, and North Carolina by the early 1950s, and in 1954 he had created two companies, Synergetics, Inc., which handled civilian contracts, and Geodesics, Inc., which took care of military and government contracts (the latter became Geometrics, Inc., in 1956). But the franchises and companies, to which Fuller typically granted nonexclusive licenses in exchange for royalties, were staffed primarily by his former students, who were not always properly credited for their often fundamental role in his design innovations. Intellectual property disputes, disappointments, and disillusionments abounded.[26]

Three Structures was no exception on this front. Fuller had taken careful measures to protect his intellectual property by applying for patents for each of the structures on display. His patent for the geodesic dome had been granted in June 1954, while his applications for the *Octet Truss* and the *Tensegrity Mast* were still pending at the time of the exhibition (the latter applied for as late as August 1959, just a month before *Three Structures* opened). In the meantime Fuller buttressed his claims to priority of invention by linguistic means, developing the art of neologism to an unprecedented degree. (In this regard Fuller's case completely substantiates a theory of invention as but an act of naming.) But Fuller's recourse to patent law and his facility for neologism did not prevent informal challenges to his assertion of intellectual property over the three structures on the MoMA lot. For example, we know from Drexler's correspondence with the two leading principals of the Raleigh branch of Synergetics—James W. Fitzgibbon and Thomas C. Howard—that the company was intimately involved with the installation of the *Octet Truss*.[27] There were those who believed that Howard should have been credited not only with the installation of the MoMA demonstration model, however, but also with its very "conception."[28] Sadao, for his part, once claimed that he was the "co-designer" (with Fuller) of the *Tensegrity Mast*, not merely its fabricator.[29] More recently, Bernie Kirschenbaum (formerly of Geometrics) has asserted that Fuller made a "grievous omission" in failing to credit Kirschenbaum and William Wainwright (a principal at Geometrics) with the MoMA radome.[30] According to Wong, most of Fuller's collaborators have claimed that he did not play a major role in the development of the radome's prototype; according to Wainwright, Fuller once even told his own patent lawyer: "I don't know if this is a Bill Wainwright radome or a Bucky Fuller radome or what it is."[31]

Whatever their individual merits, the existence of such claims foregrounds the complexity of the problem of authorship with respect to Fuller's design science. In the Far West Gallery Drexler addressed this problem by informing visitors about the vital role of Fuller's franchises, companies, and former students in both the design and execution of his inventions by including examples of their work, such as a model for an athletic center based on the principle of the *Octet Truss*, which was labeled "designed by Synergetics."[32] But Drexler's most significant contribution to the problem of authorship was the vitrine he devoted to the contributions of Kenneth Snelson. The presence and function of this vitrine within the Far West Gallery was altogether distinct from the other materials on display there. Though Snelson had been a student of Fuller's at Black Mountain College, he had never really joined the corporation. Comprising four sculptural articulations of the principle of discontinuous compression and continuous tension, the vitrine concretized in a very material way Drexler's assertion in his wall text and catalog for *Three Structures* that it was in fact the young Snelson who had discovered the principle on which Fuller's *Tensegrity Mast* was based: "The principle involved in the tension integrity mast was first discovered by . . . Snelson in 1949, following his studies at Black Mountain College with . . . Fuller. The mast in the [outdoor] exhibition is based on the same principle but employs a different configuration of parts." Further on in the same texts he referred to Snelson's discovery as "perhaps the most dramatic development to grow out of Fuller's theories," thereby deftly and tactfully acknowledging also Fuller's contribution to that discovery.[33] At the far left of the vitrine was a *Tensegrity Structure (Early Study)*—now known as *Early X Piece* (fig. 8.11)—composed of two rigid members (wooden "X" forms) that are suspended without touching one another for support in a matrix of nylon tension lines. A reconstruction of an original work dating back to 1948–49, it was this model that established Snelson's priority of invention in Drexler's mind. Adjacent to it was a model of a *Tensegrity Mast* and next to that a model of a component part of the latter. At the right was a larger model of a more complex *Tensegrity Structural System*.[34]

The museum held a press preview the day before the supplementary exhibition opened, at which it released a statement that singled out from among

Figure 8.11 Kenneth Snelson, *Early X-Piece*, 1948–49. Photo courtesy of Kenneth Snelson.

the "models illustrating various structural systems developed by Fuller and his students," one Snelson model in particular that was "similar" to the *Tensegrity Mast* in the garden but "more dramatic" in its demonstration of the tension-compression principle.[35] But despite the museum's publicity efforts, the supplementary exhibition attracted almost no media attention. The sole press references to it seem to be two tiny notices that appeared in the *New York Sunday News* and the *New York Post*, and a brief mention in a review of the outdoor exhibition in the *St. Louis Post-Dispatch*.[36] This was in sharp contrast to *Three Structures*, which was very widely reviewed in everything from art, architectural, engineering, plastics, and metal-industry magazines, to general-interest magazines, to metropolitan and local newspapers. No doubt the indoor exhibition's short run and much less spectacular appearance had something to do with this disparity. Most of the reviews of *Three Structures* marvel at the *Tensegrity Mast*—the *New York Times* art critic John Canaday calls it "the star of [the] triple show," while

Huxtable deems it "Mr. Fuller's most remarkable innovation."[37] But not a single one of the dozens of reviews or notices about the exhibition mentions Drexler's attribution to Snelson, despite the fact that this information was included in the publicity materials made available to the press at the preview for *Three Structures* (though not, it is true, in its press release per se). This omission tells us much about the celebrity function of Fuller within the media—the young Snelson was as yet unknown—and of the enduring strength of the popular myth of the lone inventor, even when all evidence is to the contrary.

Notwithstanding the silence with which Drexler's stunning revelation was greeted in the press, it was a revelation that was to have major art-historical significance. Although Snelson had continued to work with the principle of discontinuous compression and continuous tension in the decade since Black Mountain(while at the same time working as a filmmaker at the International Film Foundation in New York), he did not show his sculptural experiments publicly prior to Fuller's MoMA exhibition. Drexler's validation of the principle as Snelson's discovery, and the exposure of his work in such a high-profile forum as MoMA, proved to be extremely enabling for Snelson. As the artist himself tells it, the 1959 show was a moment of profound psychological importance, helping to redress the betrayal that he had felt ever since Fuller published the principle as implicitly his own in a 1951 issue of *Architectural Forum*.[38] In a later letter to the engineer René Motro, Snelson recalls: "That moment of recognition at the Museum of Modern Art in November 1959, transitory as it was, was quite fortifying and enabled me to once again pick up my absorbing interest in this kind of structure with the feeling that I was free and on my own."[39]

Returning to experiment again with the X-module that had been under wraps for the past decade chez Fuller (who favored instead tetrahedronal articulations of the principle), Snelson was able to create modular extension or growth "along all three axes, a true space-filling system, rather than only along a single linear axis," this last being a major limitation of the MoMA *Tensegrity Mast*.[40] Snelson applied for a patent for his discontinuous-compression and continuous-tension structure in March 1960 (which was granted in 1965), had his first solo show at the Pratt Institute in Brooklyn in 1963, and had a groundbreaking show at the Virginia Dwan Gallery in New York in 1966. In the midst of a very favorable review of the latter for the *New York Times*, Canaday remarked that "the large sculptures cry aloud for spots in any open area in the city."[41] Two years later, as it happened, Snelson mounted a spectacular installation of five superscale floating-compression structures in midtown Manhattan's Bryant Park (fig. 8.12). ("Floating compression" was Snelson's preferred term in place of "tensegrity," which he felt "always sounded . . . like some rather mealy-tasting breakfast cereal.")[42] In the present context it is hard to resist seeing Snelson's spectacle in Bryant Park as also something of a brilliant decade-in-the-making riposte to *Three Structures*. In any case, what was to be Snelson's lifelong preoccupation with floating-compression structures was now well under way. "It is necessary for me to work it out pretty much on my own," Snelson wrote to Fuller in 1972 in the midst of preparing for a series of major European shows, "[and] it's best for me to do this in the framework of the art world which [has] permitted me to squeak under the fence. It is clear this would never have happened in the Science Club."[43]

What was Fuller's response to Drexler's attribution of the discovery of the tensegrity principle to Snelson? He felt that he had been ambushed:

> You poisoned Arthur [Drexler] against me amidst a crowd at the opening of my [1959] show at the Museum of Modern Art, when I was so busy answering greetings and questions that I had no opportunity to defend myself against your utterly unexpected attack. You told Drexler that the mast I was exhibiting was yours—it was fabricated by . . . Shoji Sadao. It was not your sculpture, or a replica of your sculpture. It was the tetra-mast I had invented to show you another realization of the principle. With the crowds around me at the time, I had no time to elucidate to Arthur the facts.[44]

But Fuller fought back by other means. Shortly after *Three Structures* closed, he published his fullest exposition of the tensegrity principle to date in a 1961 article in *Portfolio & Art News Annual*.[45] Among other things, this article attempted to counter Drexler's endorsement of Snelson's assertion of his discovery. Where Fuller had formerly described his tetrahedronal mast as but a "companion" piece to geodesics, he now conflated it with the latter. The upshot of this conflation was that tensegrity was henceforth presented as a theory of geodesics. Fuller loyalists have always upheld this conflation, but various scholars have more recently argued that no such connection existed. Drawing on the testimony of an early collaborator, Duncan Stuart, Wong argues that "there is nothing immanent in the geodesic experimentation that would lead from or . . . to tensegrity. Conceptually and technically, they ensued from separate paradigms, despite sharing a common field and geometrical basis. Tensegrity as a theory of geodesic structures is, for all intents and purposes, totally fabricated and employed by Fuller to advance an appearance of cohesiveness to his life-work."[46] Indeed, in the very same 1961 article Fuller began his practice of backdating his discovery of tensegrity to 1927: "For twenty-one years before meeting Kenneth Snelson, I had been ransacking the Tensegrity concepts."[47] Fuller maintained this position for the rest of his life: In a 1982 letter to a Brian

NOVA SCOTIA

Higgins, he wrote bluntly: "Snelson did not pioneer tensegrity. I did so before [he] was born. . . . All my structures from the Dymaxion House of 1927 and all the geodesic domes are tensegrity structures."[48]

Drawing large crowds and attracting much publicity, MoMA's *Three Structures* gave top-drawer museum exposure to Fuller's visionary reach into the future, thereby laying part of the groundwork for the mainstreaming of his design science for broader public consumption during the 1960s. Organized as it was early in Drexler's tenure at the museum—he joined the department in 1956 and remained there until his death in 1987—*Three Structures* also signaled the beginning of a diversification of the museum's Department of Architecture and Design away from its historically more singularly modernist origins and preoccupations. "Being . . . out of touch with reality," as Blake once ventriloquized the critics, was henceforth wholeheartedly embraced. And if a major art-historical significance of the companion exhibition, *Buckminster Fuller: Supplementary Exhibition of Models, Photographs, and Drawings,* was that it helped to jump-start another artist's lifework by opening a psychological and professional space in which he could return to his original exploration of floating compression, on a metalevel it also affirmed the value of the material, and specifically the artistic, object as itself a site of knowledge production. In doing so, the supplementary exhibition challenged Fuller's paradoxical but deeply rooted ambivalence about the significance of the material world. The issue is not his apparently legendary lack of technical expertise but rather his assumption that the materialization of a structure or an object was essentially secondary to its ideation. Pang puts it well: As far as Fuller was concerned, his students "had only [ever] built what he had already imagined."[49] This assumption lay at the root of many of Fuller's conflicts over authorship, including that with Snelson. Fuller insisted that Snelson's *Early X Piece* was but a "special-case demonstration" of a generalized principle for which only he (Fuller) had been searching for decades. In a rambling twenty-nine-page letter of rebuttal to Snelson, which cannibalized his own publications and manuscripts in order to stake out his claim to priority, Fuller wrote: "No one else in the world but I could have seen the significance I saw" in *Early X Piece.* "I am confident that the only individual alive in 1949 who could have seen what I saw was myself."[50] In the Fuller world the polymath's prior ideation trumped the artist's discovery through making. By including *Early X Piece* in the 1959 supplementary exhibition, Arthur Drexler turned this most Fullerian assumption completely on its head.

Figure 8.12
(opposite) View of Kenneth Snelson's exhibition in Bryant Park, New York, 1968.
Photo by Kenneth Snelson. Used by permission

9

R. Buckminster Fuller

A Technocrat for the Counterculture

Fred Turner

In 1965 R. Buckminster Fuller was seventy years old. Short, plump, bespectacled, and, when he spoke in public, often clad in a three-piece suit with an honorary Phi Beta Kappa key dangling at his waist, Fuller looked like nothing so much as an early twentieth-century plutocrat. When he took the stage, he filled the air with hours of technocratic talk, much of it of his own design. Industry! Technology! The Space Program! Leaping from topic to topic across sentences decorated with his own fabulously recondite vocabulary, Fuller spun a cotton-candy of machine-age dreams. New chemicals, new alloys, and new ways of measuring the ever-more massive output of international industry—like the most visionary corporate executive of the high industrial era, Fuller urged his listeners to imagine a world made good by machinery, management, and design.

Yet for all his obvious allegiance to the ideals of the industrial world, Fuller was also a hero to the young members of the American counterculture. Two of his books—*Ideas and Integrities* (1963) and *Operating Manual for Spaceship Earth* (1969)—became staples of hippie libraries across America. His lectures became magnets for the young, and his geodesic domes became the preferred housing of many rural communards. In 1968 his writings became the inspiration for the publication that has long been seen as the Bible of the back-to-the-land movement and a signal document of the counterculture, the *Whole Earth Catalog.* To all those who had wandered off into the plains of Colorado and the hills of New Mexico to build new communities, and to all those who dreamed of making such a move, R. Buckminster Fuller was an inspiration.

But why? What was it about this aging designer and engineer that made him so attractive to a movement that ostensibly rejected industry, technology, and the advice of anyone over thirty years old?

To answer these questions requires cutting both the American counterculture and R. Buckminster Fuller free from the tangle of myths that have grown up around them. Since the late 1960s, scholars and journalists alike have tended to read the American counterculture in terms set by its proponents at the time. Then and now, analysts have argued that the counterculture represented a collective turn away from the technologies and organizational forms of cold war America. Likewise, thanks in part to his own ability to turn his own life into compelling copy, R. Buckminster Fuller has often been depicted as a sui generis genius, a tinkering autodidact in the tradition of Thomas Edison and Alexander Graham Bell. Yet, despite their respective claims to the contrary, neither Fuller nor the American counterculture emerged entirely outside the orbit of the era's military-industrial complex. On the contrary, Fuller and his theory of "comprehensive design" offered many in the 1960s a way to embrace the technologies, the technocratic politics and the flexible, collaborative work styles of the cold war military and industrial worlds even as they built their own alternative communities.

Between Nuclear Holocaust and Consumer Cornucopia

To make sense of the countercultural turn toward technology and Fuller, we need first to remember that Americans in the 1950s, and especially American children, lived under the imminent threat of nuclear Armageddon. In 1967 social psychologist Kenneth Keniston interviewed a group of young men and women who had taken part in a series of anti–Vietnam War efforts. Hoping to uncover the roots of their activism, he asked them to recall their earliest memories. One young woman described the day an encyclopedia salesman sold her mother volume A of the *Encyclopedia Britannica*: "I remember reading it and seeing a picture of an atomic bomb and a tank going over some rubble. And I think I became hysterical. I screamed and screamed and screamed."[1] This young woman was hardly alone. For those who were children at the height of the cold war, the possibility of a nuclear holocaust felt very real. In elementary school they had been taught to "duck and cover" under their desks if they should happen to see a nuclear flash. They had been shown government-sponsored films in which children their own age sprinted through neighborhoods that had been reduced to atomic rubble, hunting for the local fallout shelter.[2] Ever since the Soviet Union had first tested an atomic bomb in 1949, Americans had suffered under a thick cloud of nuclear anxiety. Would the devastation that the *Enola Gay* had wreaked on Hiroshima somehow visit American cities? Would New York someday look like Nagasaki?

By the late 1950s, many Americans had begun to fear that the bomb had become a way of life. The military agenda of the nation at the time seemed to lock adults into a particularly constrained way of life, a way of life that the youth of America would presumably be forced to lead as soon as they grew up. As Elaine Tyler May has pointed out, the dominant social style of the middle and upper classes during the postwar years could be described as "containment."[3] Much as military and government planners sought to "contain" communism, the men and women of middle America sought to constrain their emotions, maintain their marriages, and build safe, secure, and independent homes. Like the Air Force soldiers who scanned America's borders for incoming Soviet bombers, many Americans took to monitoring the boundaries of their own lives.

Containment was the order of the day in the workplace as well. For critics on the left in particular, society seemed to be increasingly dominated by pyramidal organizations run by buttoned-down, psychologically fragmented men. "As the means of information and of power are centralized," wrote the sociologist C. Wright Mills in 1956, "some men come to occupy positions in American society from which they can look down upon . . . and by their decisions mightily affect, the everyday worlds of ordinary men and women."[4] Under the controlling eye of this "power elite," Mills argued, ordinary Americans found themselves trapped in corridors and offices, unable to envision, let alone take charge of, the entirety of their work or their lives. Ordinary people lacked the ability to "reason about the great structures—rational and irrational—of which their milieux are subordinate parts," he explained.[5] So, too, in a way, did the men at the top. For critics like Mills, both the masters of bureaucracy and their minions suffered from a paring away of emotional life and a careful separation of psychological functions. In the wake of World War II, wrote Mills, rationalization had begun to give rise to "the man who is 'with' rationality but without reason, who is increasingly self-rationalized and also increasingly uneasy."[6] This man, wrote Mills, was a "Cheerful Robot."[7]

Alongside the twin threats of the bomb and of a stultifying, mechanical adulthood, however, the young Americans of the 1960s also enjoyed an unparalleled level of affluence and, with it, a cornucopia of consumer goods. At one level these goods, too, were part of America's cold war military tool kit. In 1959, for instance, Vice President Richard Nixon found himself facing down Soviet premier Nikita Khrushchev in a model kitchen at the American Exhibition in Moscow. Nixon proudly told the scowling Khrushchev, "There are 44 million families in the United States. . . . Thirty-one million families own their own homes and the land on which they are built. America's 44 million families own a total of 56 million cars, 50 million television sets and 143 million radio sets. And they buy an average of nine dresses and suits and 14 pairs of shoes per family per year."[8]

Yet, for the children of the 1950s who would become the rebels of the 1960s, cars, TV sets, and radios also offered an escape from the shadows of the cold war. Teenagers found themselves surrounded by appliances and automobiles and opportunities for education and employment that their parents, growing up in the Depression, could hardly have imagined. As many commentators remarked at the time, this affluence transformed adolescence into a true interregnum between the freedom of childhood and the employment and family demands of adulthood.[9] For the ever-increasing numbers of middle- and upper-class youths in particular, adolescence became a time for personal exploration.

By the late 1960s, then, young Americans confronted a dilemma. On the one hand, the world of military and industrial bureaucracy and the technologies associated with it threatened to end their lives, either by destroying the earth in a nuclear holocaust or by demanding that as they became adults, young men and women transform themselves into "Cheerful Robots." On the other hand, however, that same bureaucracy had endowed their lives with all sorts of technologically supported pleasures, including music, television, and travel. Moreover, thanks to the power of postwar industry, the college graduates of the 1960s would have no trouble finding jobs. But would those jobs provide the same sorts of satisfactions that adolescence had offered? Many had their doubts. "There are models of marriage and adult life, but . . . they don't work," recalled the same young woman who had discovered the atom bomb in the encyclopedia. "There is that whole conflict about being professional, leading a middle-class life which none of us have been able really to resolve. How do you be an adult in this world?"[10]

Comprehensive Design as a Way of Life

It was with this question in mind that many turned to R. Buckminster Fuller. If the politicians and CEOs of mainstream America were distant and emotionally reserved, Fuller was playful and engaged. And like his young audiences, he displayed a highly individualistic turn of mind and a deep concern with the fate of the species. But it was not simply Fuller's character that drew his audiences to him. Rather, it was the resolution he offered to the paradox that confronted the young adults of the 1960s. As he moved from university to university, collaborating with college students, giving speeches, and designing new technologies, Fuller exemplified a way of making a living alongside the academy and industry without becoming in any way a bureaucrat. Moreover, his rhetoric and his theories of technology seemed to integrate the most microcosmic aspects of daily life and the most macrocosmic forces shaping human survival. For Fuller, design could be more than a stage of manufacture associated with cold war industry; it could be a world-saving way of life.

In a 1949 essay that he later expanded and reprinted in *Ideas and Integrities*, a volume that circulated throughout the counterculture, Fuller codified that vision as an expression of his own professional goals and beyond them, of a new professional category, the "comprehensive designer."[11] In *Ideas and Integrities* Fuller located the origin of his vision squarely in his personal experience. During World War I, he wrote, he had watched his four-year-old daughter, Alexandra, die of infantile paralysis, in part, he believed, because the family's home was badly built.[12] At the time, he was working as a contractor with the Navy. As a former junior officer, he had seen how, with proper coordination, extraordinary industrial resources could be mustered to solve military problems. In his view his daughter had died directly from a disease but indirectly from a failure to distribute the world's resources appropriately.[13] This conviction grew during World War II and the early years of the cold war, where once again Fuller saw the full scope of industrial production at work, as well as the inequality with which those resources were distributed. In Fuller's view the natural world was governed by a series of laws that kept it in harmonious balance. In his experience, however, the mid-twentieth-century social world was one in which the material goods created in accordance with those laws were not being evenly distributed and where children were dying as a result. Politicians, generals, corporate leaders—each put the needs of his own organization first when it came to resources. What humankind required, he argued, was an individual who could recognize the universal patterns inherent in nature, design new technologies in accord with both these patterns and existing industrial resources, and see that those new technologies were deployed in everyday life.

This individual he explained would be a "comprehensive designer."[14] According to Fuller, the comprehensive designer would not be another specialist, but would instead stand outside the halls of industry and science, processing the information they produced, observing the technologies they developed, and translating both into tools for human happiness. Unlike specialists, the comprehensive designer would be aware of the system's need for balance and the current deployment of its resources. He would then act as a "harvester of the potentials of the realm," gathering up the products and techniques of industry and redistributing them in accord with the systemic patterns that only he and other comprehensivists could perceive.[15] To do this work, the designer would need to be able to access all of the information generated within America's burgeoning military-industrial bureaucracy yet at the same time remain outside it. He would need to become "an emerging synthesis of artist, inventor, mechanic, objective economist and evolutionary strategist."[16] Constantly poring over the population surveys, resource analyses, and technical reports produced by states and industries, but never letting himself become a full-time employee of any of these, the comprehensive designer would finally see what the bureaucrat could not: the whole picture.

This vision would allow to him to realign both his individual psyche and the deployment of political power with the laws of nature. If, as so many in the 1960s had begun to suspect, the bureaucrat had been psychologically broken down by the demands of his work, the comprehensive designer would become whole again. Neither engineer nor artist, but always both simultaneously, he would achieve psychological integration even while working with the products of technocracy. Likewise, where bureaucrats applied their power by means of political parties and armies, and in Fuller's view, thus failed to properly distribute the world's resources, the comprehensive designer would apply power systemically. That is, he would analyze the data he had gathered and attempt to visualize the world's needs, now and in the future. He would then design technologies that would meet those needs. The technologies would so reshape the environment within which people worked as to reorganize society itself. This new society would see its resources distributed not in keeping with the demands of politicians but with the natural laws that already kept the world system of nature in balance. Agonistic politics, Fuller implied, would become irrelevant. What would change the world was "comprehensive anticipatory design science."[17]

With the notion of comprehensive design, Fuller offered his readers a way to embrace the pleasures and power associated with the products of cold war industry even as they avoided becoming bureaucratic drones. Moreover, Fuller implied that the reshaping of the individual life and its reorientation around principles of comprehensive design could save not only the individual but the species. As he put it in *Ideas and Integrities*: "If man is to continue as a successful pattern-complex function in universal evolution, it will be because the next decades will have witnessed the artist-scientist's spontaneous seizure of the prime design responsibility and his successful conversion of the total capability of tool-augmented man from killingry [*sic*] to advanced livingry [*sic*]—adequate for all humanity."[18] In Fuller's view the comprehensive designer not only didn't need to don a gray flannel suit when he went to work—he actually needed to become an artist and an intellectual migrant. To a generation preoccupied with the fear of becoming lock-step corporate adults, R. Buckminster Fuller offered a marvelously playful alternative, but one that was not mere play. It was a way to preserve the human future.

Despite Fuller's claims to have coined the term in response to his unique biographical conditions—a claim that reinforced the notion that his own life should serve as an example for his readers—Fuller's vision of the comprehensive designer carried with it intellectual frameworks and social ideals formulated at the core of military research culture. Foremost among these was Fuller's notion of the world as an information system. In his numerous autobiographical writings Fuller traced the origins of his ideas about the world as a system to his great aunt Margaret Fuller's involvement with the transcendentalists and especially to his time on board ships—which he considered closed systems—when he was a naval officer.[19] Yet his writings also bear the imprint of cold war–era, military-industrial information theory. For Fuller, as for the information theorists of World War II and the systems analysts of later decades, the material world consisted of information patterns made manifest. These patterns could be modeled and manipulated by information technologies, notably the computer. The computer in turn could suffice as a model for the human being.[20] After all, while Fuller's comprehensive designer promises to be psychologically integrated as specialists are not, that integration depends on the designer's ability to process vast quantities of information so as to perceive social and technological patterns. Fuller's comprehensive designer is, from a functional point of view at least, an information processor, and as such, as much a descendent of cold war psychology and systems theory as a child of Fuller's own imagination.[21]

Even Fuller's seemingly unique work style echoes the collaborative ethos of World War II research. According to Fuller and, later, to his countercultural admirers, the comprehensive designer came by his comprehensive viewpoint only by stepping away from the industrial and military institutions in which specialists had long been trapped. Only the freestanding individual "could find the time to think in a cosmically adequate manner," he explained.[22] By scanning the horizon of specialties and moving from institution to institution, Fuller argued, the comprehensive designer could glean enough information to see the entire "system." Fuller himself lived according to this ethos: for most of his career he migrated among a series of universities and colleges, designing projects, collaborating with students and faculty—and always claiming the rights to whatever these collaborations produced. By the early 1960s, Fuller was traveling more than two-thirds of every year.[23] In his writings Fuller offered his travels as a model of the proper behavior for a comprehensive designer and suggested that such a life was genuinely new. Yet a quick glance back at the laboratories of Los Alamos or MIT's Rad Lab during World War II would have reminded Fuller's audiences that interdisciplinary migration and multi-institutional collaboration were key features of the military research world. They were, in fact, the social processes for which cybernetics and systems theory had served as a universal discourse.[24] Even as Fuller claimed to be a sui generis intellectual, and even as his audiences celebrated his ideas and his lifestyle as harbingers of the future, Fuller's allegiance to systems theoretical perspectives, his faith in information as the substrate to experience, and his collaborative work style all carried with them links to the very military-industrial complex that the youth movements of the 1960s claimed to want to overthrow.

Comprehensive Design and the Politics of Consciousness

Yet, strangely enough, it was these links that helped make Fuller so attractive to so many at the time. Today, Americans often remember the youth movements of the 1960s as a single mass attack on institutions and cultural styles of cold war America. However, while they did share aversions to the Bomb and to the suburbs, members of those movements tended to adopt one of two quite different postures toward social change. In the early 1960s, alongside the civil rights movement in the South and the free speech movement at Berkeley, students began to organize into a political movement that would become known as the New Left.[25] For these activists the key to social change lay in political action. Accordingly, its members formed new parties (such as Students for a Democratic Society, or SDS), staged conventions, issued manifestos and marched against the Vietnam War. If elements within the New Left began to experience forms of solidarity like those they helped to build into the world outside the movement, they did so as an aftereffect of their own organizing. Within the New Left, true community and the end of alienation were usually thought to be the result of political activity, rather than a form of politics in its own right.

The reverse was true among what I will call the New Communalist wing of the counterculture.[26] If the New Left had grown up out of cold war social struggle, the first stirrings of New Communalism appeared within the artistic bohemias of cold war Manhattan and San Francisco, among the peripatetic Beats, and finally, among the mystics and acid heads of the San Francisco Bay Area in the early 1960s. For the New Communalists the key to social change was not politics but mind. In the 1969 volume that first popularized the phrase "counter culture," Theodore Roszak spoke for many New Communalists when he argued that the central problem underlying the rationalized bureaucracy of the cold war was not political structure but the "myth of objective consciousness."[27] This state of mind, wrote Roszak, emerged among the experts who dominated rationalized organizations and was conducive to alienation, hierarchy, and a mechanistic view of social life. Its emblems were the clock and the computer, its apogee "the scientific world view, with its entrenched commitment to an egocentric and cerebral mode of consciousness."[28] Against this mode Roszak and others proposed a return to transcendence and with it, a simultaneous transformation of the individual self and its relations with others:

> This . . . is the primary project of our counter culture: to proclaim a new heaven and a new earth so vast, so marvelous that the inordinate claims of technical expertise must of necessity withdraw in the presence of such splendor to subordinate and marginal status in the lives of men. To create and broadcast such a consciousness

of life entails nothing less than the willingness to open ourselves to the visionary imagination on its own terms.[29]

In the mid-1960s this new consciousness became the basis of the largest wave of communalization in American history. In the two centuries before 1965, historians and sociologists have estimated that Americans established more than six hundred communes.[30] Between 1965 and 1972 journalists and sociologists have estimated that somewhere between two thousand and six thousand communes were created, with most appearing between 1967 and 1970.[31] Virtually all of these communities were built by young, white, middle- and upper-class youths, and with few exceptions, they had little to do with the New Left. Rather, the communards of the late 1960s aimed to organize themselves around the pursuit of a shared consciousness and with it, a leveled social structure that would obviate the need for conventional politics. One of the earliest such communes, Drop City, blossomed in a cluster of geodesic domes on the plains of Colorado in 1965 (fig. 9.1).[32] As cofounder Peter Douthit, better known as "Peter Rabbit," explained at the time: "There is no political structure in Drop City. Things work out; the cosmic forces mesh with people in a strange complex intuitive interaction. . . . When things are done the slow intuitive way

Figure 9.1 Two domes at Drop City soon after construction. © Estate of R. Buckminster Fuller. All rights reserved. Used by permission. Source: Special Collections, Stanford University Libraries.

the tribe makes sense."[33] At Drop City individuals were free to come and go whenever they liked and to pursue what interested them moment to moment. This freedom they believed would lead to a greater state of collective harmony, with one another and with unseen forces in the universe. "We dance the joy-dance [*sic*], we listen to the eternal rhythm, our feet move to unity . . . live-love-joy-energy are one," wrote Rabbit. "We are all one."[34]

For the Droppers, as for thousands of other young communards, consciousness formed the foundation of a new kind of sociability—holistic, collaborative, antibureaucratic. Small-scale technologies in turn opened the doors to consciousness and, thus, to this new social world. LSD, water pipes, stereo gear, books such as the *I Ching*, Norbert Wiener's *Cybernetics*, and especially, the writings of R. Buckminster Fuller—for the New Communalists, each of these items served as a tool with which to remake the self and, with it, the group. They also served as bridges between the industrial world that the New Communalists had left behind and the postindustrial future they hoped to build. Fuller had patented the geodesic dome, for instance, in 1951; between 1954 and 1957 the American military deployed hundreds of these domes to house radar installations across a three-thousand-mile early warning line built in Canada.[35] During those same years, Fuller's domes were exhibited worldwide at trade fairs and expositions as evidence of American technological ingenuity. Yet even though they had served as emblems of America's military-industrial might, at Drop City they also became emblems of an America transformed. The multicolored panels of the geodesic domes at Drop City for instance, were made from the roofs of junked automobiles. The commune's long-haired founders had spent days chopping the roofs out of old cars with hand axes and electric saws and then bolting them to wooden frames. In the process they turned an industrial artifact into an occasion for hand craft and collective labor. The houses they built in turn became emblems of a new mind-set. As one Drop City resident put it, "The domes have a sort of cosmic guidance. All those triangle sections coming together to make a single dome, a self-supporting thing. It's like a community can be."[36]

In that sense the builders of Drop City's domes had become comprehensive designers. As they chopped up the roofs of old cars and bolted them together into complex geometric patterns, the communards of the back-to-the-land movement embraced the intellectual and material output of American industry, as well as the collaborative, freelance work styles of military-industrial research. At the same time, they disassociated themselves from the Bomb and the bureaucratic professional culture that they imagined had produced it. In this way they both rejected their parents' world and, ultimately, found a way to make their own place in it.

The New Communalists also set a Fulleresque example for a generation of young Americans. In 1968, San Francisco–based multimedia artist and entrepreneur Stewart Brand and his wife, Lois, published a sixty-one-page guide to books, mechanical devices, and outdoor gear that they hoped would be useful to those heading back to the land, the *Whole Earth Catalog* (fig. 9.2). Over the next four years the *Catalog* would grow to more than four hundred pages, would sell more than a million copies, and would win the National Book Award. To some who lived on the land, and to many who didn't, the *Catalog* became a primer in comprehensive design. As Brand put it in his introduction to the *Catalog*'s first section, "Understanding Whole Systems," "The insights of R. Buckminster Fuller initiated this catalog."[37] Sized somewhere between a tabloid newspaper

Figure 9.2
The cover of the first *Whole Earth Catalog* (1968).
Photo courtesy of Stewart Brand. Source: Special Collections, Stanford University Libraries.

and a glossy magazine, the *Whole Earth Catalog*, like Fuller's own writings, offered readers a vision of technology as a means by which to escape industrial bureaucracy while living synergistically off its fruits. Consider the *Catalog*'s opening statement. On the inside cover of every edition, Stewart Brand defined the *Catalog*'s "PURPOSE":

> We *are* as gods and might as well get good at it. So far, remotely done power and glory—as via government, big business, formal education, church—has succeeded to the point where gross defects obscure actual gains. In response to this dilemma and to these gains a realm of intimate, personal power is developing—power of the individual to conduct his own education, find his own inspiration, shape his own environment, and share his adventure with whoever is interested. Tools that aid this process are sought and promoted by the WHOLE EARTH CATALOG.

Brand's definition clearly states the countercultural critique of hierarchical, establishment institutions as emotionally and geographically remote from the lives of citizens and, on the whole, destructive. At the same time, he intimates that he and the reader are like gods in at least two senses, one local and one global, and both familiar from Fuller's *Ideas and Integrities.* On the local level the individual reader is like a god in that each person has the power to conduct life as he or she wishes, as long as he or she can find the appropriate tools. For Brand, as for Fuller, the system of the universe is complete—it is not something we can put together but something that is together in its own right. At the local level our job is to turn its energies and resources to our own purposes. In keeping with the countercultural critique of bureaucracy, we must pursue our own individual transformation and, with it, the transformation of the world.

These transformations depend, however, on our understanding the world as a system of invisible forces. At the global level, like Fuller's comprehensive designer or perhaps a cold war systems analyst, Brand's reader enjoyed the power of a god to survey the whole earth below him. The front cover of many editions of the *Whole Earth Catalog* featured an image of the earth seen from space. Simply by picking up the *Catalog*, the reader became a visionary of a sort. This vision, though, had been made possible by the cameras of NASA and, more generally, by the fact that the reader was a member of the most technologically advanced generation on Earth. In the *Whole Earth Catalog* the same technocracy that had spawned the world of the white-collar worker and the war in Vietnam had granted those who rejected both the power to see the world in which they lived as a single whole.

In this sense the *Catalog* suggested that its readers could become comprehensive designers as they read. As soon as they opened the book, their eyes could roam across what looked to be a whole planet's worth of goods: books, tepees, handsaws, radios, motorcycles—you name it. Simply by thumbing through the *Catalog*, readers could imagine themselves as masters of a universe of information and designers of their own lives. The *Catalog*'s offerings served in turn as tools with which the reader could deploy the principles of comprehensive design in everyday life. In the pages of the *Catalog*, as on the rural communes it was created to serve, a backpack or a tent did not simply offer a means of escape into the woods. It offered readers a chance to join an invisible community of nomads, to act in accord with the ancient energies of nature and to become a more "whole" person in the process. That is, these goods would help transport the reader into an environment in which he or she might be able, at the global level, to spot and, at the local, personal level, to act in accord with, the laws of nature. In that way the *Catalog*'s small-scale technologies, its backpacks and tents, and, of course, geodesic domes—a staple of the *Catalog*, as well as of many communes—were not so much tools for action as tools for vision. They offered readers the means to transform the products of high-technology industry into a way of seeing the world as a whole. Having grasped that vision, these comprehensive designers could create new communal worlds of their own and by their example, individual and collective, save the world as a whole from the perils of bureaucratized industry.

Conclusion

For the young of the 1960s the logic of comprehensive design embodied a dizzying set of analogies that placed their own lives at the center of the universe. The individual life, the new community, the world as a whole—as glimpsed in the pages of the *Whole Earth Catalog* or lived on a communal farm, each was an emblem of one another and all constituted an indissoluble whole. Equipped with the proper tools, the young American could scan the whole globe, perceive its hidden patterns, and act in his or her own—and, presumably, the world's—best interests. If the cold war bureaucrat sat huddled in his office, glimpsing only the most partial fragments of the human enterprise, the comprehensive designers of the back-to-the-land movement positioned themselves at the fringes of American society and thereby sought to take in a wider view. Having forsaken the bureaucratic towers of technocracy, they could take up its many technological products and turn them to a new end: the transformation of the individual consciousness and with it the founding of a new society. At the same time, they could escape the conundrums of adulthood that beset their generation. After all, what could be more important or more fun than building a new society?

R. Buckminster Fuller's own life seemed to prove the point. As they read his books and flocked to his lectures, many middle- and upper-class youths hoped to harness the economic power of American industry to build independent, flexible lives and to grow up to enjoy them as much as he did his. Yet, to the extent that they tried to build those lives on communes, most failed. By the mid-1970s virtually all of the communes built over the preceding decade had disappeared. While the vision of communities founded on shared consciousness alone held enormous appeal in theory, it crumbled in the face of the material realities of rural farming and the complexities of collective life.

For the young of the late 1960s, R. Buckminster Fuller's vision of comprehensive design had seemed to offer an escape from the need to enter institutions, to confront other individuals, to struggle over the distribution of resources and the proper organization of life. In the coming years Fuller's hope for a world of individuals equipped with vast databases of information and the capacity to see—and manage—the world as a whole would animate the rise of the personal computer and the introduction of the Internet. Yet, even as the theory of comprehensive design has lingered in the cultural atmosphere and, with it, the hope for a social life built on interpersonal harmony, free commerce, and a lack of bureaucracy, so too has the failure of the communes. In 1973 the founders of Drop City sold the commune's land and left their geodesic domes to collapse under the incessant Colorado winds.[38] As they had learned, tools alone could not sustain community, nor could careful attention to design replace the nitty-gritty, everyday work of politics.

10 Fluid Geographies

Politics and the Revolution by Design

Felicity D. Scott

World Thinking, Thinking World

In an "Intermedia" column published April 3, 1970, in the *Los Angeles Free Press*, Gene Youngblood narrated a recent encounter with R. Buckminster Fuller. Entitled "Earth Nova," the article was the first of a series appearing in this underground paper that the young critic would dedicate, enthusiastically, to promoting Fuller's "revolution by design." Youngblood had been awoken from "a dream of crystal ships" at 8:00 a.m. by Tom Turner, director of research for the World Game, and summoned to a World Game simulation demonstration about to take place at Southern Illinois University at Carbondale, home of the World Resources Inventory. Ten minutes later TWA called to say tickets were waiting. Six hours later Youngblood found himself in St. Louis, Missouri, connecting with Fuller, who was arriving from Atlanta, and Michael Binelli, a geodesic dome building activist coming in from New York. The three then took off in a small plane on a rather treacherous journey to Carbondale, 150 miles away. As Youngblood recalled: "The red glow of the instrument panel bathed our faces eerily, and I realized we were flying totally blind, with visibility no farther than the tip of the wings."

Faced with a sense of imminent danger, Youngblood took solace in Fuller's faith in technology. "I was thinking how appropriate it was to be with Bucky this way, trusting our lives to the very design integrity that he has spent a lifetime defending as humanity's only hope for success."[1] Youngblood's invocation of the trope of the pilot "flying blind" was not incidental to this story: not only had it served in Fuller's *Utopia or Oblivion: The Prospects for Humanity* (1969) to indicate the inherent limitations of "physically conceptual models" to represent invisible energy forces,[2] but it was often used to demonstrate the mutual interrelationship between computerization and the deracinated human condition Fuller sought, that "world-around freedom of living" motivating his World Game. "World-around jet travelers now commit their lives to the computer's reliability," Fuller explained in justification of the World Game's cybernetic infrastructure.[3]

The World Game was in many ways a continuation of Fuller's long-standing project of redirecting "weaponry arts" for "livingry arts."[4] The military model of a computerized, multiplayer logistical game—the War Game—had here been transposed into a game in which competing teams of players would "each develop their own theory of how to make the world work successfully for all of humanity."[5] As Fuller recalled in "How It Came About (World Game)," the project was initially proposed to the United States Information Agency to serve as an exhibition at Expo 67 in Montreal. With the war in Vietnam very much on his mind, the project's dedication to the question of "How to Make the World Work" through equitable distribution of global resources aimed to counteract what Fuller regarded as the likelihood of the rest of the world's plummeting opinion of the nation.

Visitors to World Game, as conceived for Expo 67, would enter a massive, four-hundred-foot-diameter, five-eighths-sphere dome. Inside this dome was suspended a one-hundred-foot-diameter world globe that would periodically transform into an icosahedron before flattening out onto the floor in the manner of Fuller's Dymaxion air-ocean world map, a limited distortion cartographic projection of the earth. The "great map," Fuller explained, "would be wired throughout so that minibulbs, installed all over its surface, could be lighted by the computer at appropriate points to show various, accurately positioned, proportional data regarding world conditions, events and resources."[6] Looking down from a high balcony onto this football field–sized, computer-driven Dymaxion data map, with information continuously streaming from massive data banks and satellites, "major world individuals and teams" chosen, as Fuller explained, for their "lack of bias as well as for their forward-looking competence," would pursue "grand world strategies."[7] They would develop competing scenarios in real time for distributing resources to a global humanity liberated from nation-based constraints, a humanity comprising nomadic "citizens of the world."

World Game was not realized at Expo 67, but the spectacular 250-foot diameter, three-quarters sphere, transparently clad geodesic dome Fuller designed for the occasion would help catapult him to the height of his worldwide fame in the late 1960s. His patented invention, the geodesic dome, had of course been adopted years earlier by the State Department as a pavilion typology to represent the country at world's fairs and prior to that by the U.S. military in a limited fashion for use as rapid deployment structures. Yet although projects like the 1959 pavilion in Moscow had gained significant publicity, it was the enormous media coverage of the American Pavilion in Montreal that transformed his status from obscure and eccentric inventor to a household name in American ingenuity.[8] And somewhat paradoxically—given the distinctly mainstream character of world's fairs and their role in cold war propaganda, as well as his connections to the U.S. military—Fuller was also at this moment being embraced by the American counterculture and radical movements as something of a prophet for peace, equality, whole earth ideology, and transcendental well-being. Geodesic domes had, moreover, become an architecture of choice for the counterculture. Understood to be an ecologically sound technology for housing a postrevolutionary society, Fuller domes seemed to offer a means for escaping the commodified lifestyle and normative values of mainstream America.[9]

On the one hand, it is not difficult to understand this attraction; Fuller raised issues that were on many people's minds, and his ideas seemed to flow amorphously into concerns then being voiced by the nation's youth. Appealing at once to a growing environmental consciousness and sense of apocalyptic doom, and their counterpart in an escalating survivalist ideology, he captivated his audience by posing questions such as: "Is our Spaceship Earth's biosphere to be [an] omnihumanity-sustaining environment or an omnilethal one?"[10] But perhaps more important, he offered answers, cast in a suitably spaced-out, seemingly metaphysically happening register. It was a reception that he assiduously cultivated both through extensive lecture tours and interviews in the underground press, as well as through strategic alliances with writers like Youngblood. Appearing in the *LA Free Press* alongside articles on Kent State, the war in Vietnam and its extension into Cambodia, radical ecology, Charles Manson, the Black Panthers, gay rights struggles, and the women's movement, Youngblood's series of World Game articles do, on the other hand, seem in retrospect distinctly out of place in this otherwise highly politicized context.[11] For driving Fuller's one-world vision was, as I want to trace here, an avowed and programmatic rejection of politics.

Tomás Maldonado noted at the time that Fuller's "technocratic utopianism" not only disarticulated the social from the technological but sought the eradication of the very domain of the political in human relations. "In our view," he wrote, correcting this lapse, "the 'Revolution by Design' should be the result both of the technical imagination, and of what C. Wright Mills called the 'sociological imagination'—both technical courage and social and political courage."[12] To Fuller's thinking, however, politics, all too broadly defined, was the underlying cause of poverty, hunger, illness, inequality, immobility, and warfare, and its overcoming through his revolution by design (as demonstrated by World Game) was to launch humanity into a peaceful global condition. Without political parties or ideologies vying for control of state power, and without nation states vying for domination over world resources, warfare, Fuller argued, would simply become redundant. At stake, then, in revisiting this moment of Fuller's reception is not so much the question of the appropriateness of this mutual identification of Fuller and the counterculture but its effects on the political contours of the late 1960s and early 1970s.

Free Press

Youngblood had joined Fuller in the World Game crusade, situating his own work as part of the effort "to get World Game to the people of the world through the global intermedia network." If Fuller himself was shortly to address the United Nations in New York with a World Game demonstration (one aiming, of course, to undermine the international politics on which the United Nations was premised), Youngblood situated his own writings, both for the mass media (he listed *Look*, *Harper's*, *Atlantic*) and the underground press as part of this necessary "information process."[13] What Youngblood termed the "videosphere" or global television was central to World Game's desired project of sponsoring "a global climate of outrage against the nation state system."[14] Lecturing to architecture students at the University of Southern California on the occasion of the First Earth Day (April 22, 1970), he made this connection explicit:

> Since, with the videosphere, you have the ability to address the entire world with information, and since the media networks are inherently information-hungry, news hungry, [they] can't stand it if they don't have something new to say on the Six O'Clock News each day—obviously World Game is going to be irresistible to them. When you prove scientifically that the existence of sovereign nation states is responsible for the destruction of the planet, this is going to be like catnip to the news media. They're like junkies. They need media fodder to sustain themselves. And they have no integrity.[15]

The process would thus be virtually automatic, driven by the inherent logic of the media networks themselves.

During the flight to Carbondale Fuller proclaimed, as he had done repeatedly over the previous half-decade: "Something very important is happening around the world. . . . I can sense it. Young people really are beginning to understand what's going on." The two young men assured him that his hope was not misplaced: "Everywhere we go they want to know about Fuller, and about World Game." But something else was clearly on Fuller's mind. Youngblood went on to recount that in Oxford, only a week earlier, Fuller "had encountered his first anti-Fuller propaganda: a poster depicting the top of his head as a Geodesic dome with marines perched in it." Youngblood stepped in to defend Fuller: "Obviously," he retorted, "a petty reference to military use of Geodesic structures," adding, parenthetically: "The fact that in 1954 the Marines, who ought to know, pronounced air-deliverable domes 'the first basic improvement in mobile environment controls in 2600 years' seemed to be of little interest to the anti-Fuller faction." If those present had appeared shocked by the blasphemous image, Fuller insisted that he hadn't been, responding with the remark that "Marines eat rice too, you know."[16]

The poster was most likely the "UN Official Program," "Welcome to the Buckminster Führer Show," produced in 1968 by Arse in London;[17] if not this artifact, then it would have been something similar.[18] Fuller's quip—that "Marines eat rice too"—was not simply a universalizing pun (in somewhat poor taste) involving a prime staple of the Southeast Asian diet while war raged in Vietnam. Rather, it performed precisely the displacement at the center of his revolution by design: a shift from questions of politics to those of world resources. Fuller repeatedly insisted that were it not for politics, the utopian prospects harbored in "doing-more-with-less" might become reality: "Total success for all, can be realized only through a design-science revolution of university aged youth. This revolution is trying to articulate itself everywhere, but it gets bogged down by political exploiters."[19] The residual role afforded to politics was that of "a secondary service—a stewardess function—of polite supervision of the passengers' 'adjusting of their seatbelts' for the great world 'takeoff' for physical success of all mankind."[20]

Youngblood's retort was hardly unexpected. Not only had he firmly aligned himself with Fuller's philosophy—even inviting him to contribute the introduction to his own book, *Expanded Cinema*, of 1970—but he often ventriloquized these key aspects of Fuller's thought: the importance of youth and the rejection of politics (both with respect to the political life of sovereign nation-states and revolutionary struggles).[21] In Youngblood's words:

> Although we claim to be the seeds of a new human consciousness, still we looked to the past, seeking an answer in politics. Students for a Democratic Society drew Marx's conclusions on the wall, forgetting that their politics could lead to democratic suicide. As we emerge into the planetary society of the future, the geopolitics of the past are the most dangerously constraining myths of our present. But we wanted to repeat history. We wanted another French Revolution, we wanted another Russian Revolution, we wanted another revolution of the sort that produced the America we now reject.[22]

Dismissing revolutionary ideals, Youngblood, too, had come to believe that evolution through technology would replace revolutionary politics as a mechanism of social change. "The current popularity of revolution," he noted, marking his distance from "clenched fists, or the Rolling Stones or the SDS," "is a revolutionary posture not a liberal one." Rehearsing the ideology of the new, he announced: "Evolution never repeats history; evolution is always at the frontier."[23] Fuller had clearly found in Youngblood a reliable agent operating from within the counterculture, one dedicated to continuing his design-science message into the 1970s.

Coupled with this rejection of politics were, as already indicated, geopolitical claims that were equally founded on a generational argument, as well as on a slippage between political revolutions and those in the technological domain. To the agricultural, scientific, and industrial revolution Youngblood proposed "to add a fourth fundamental transformation in the behavior of man on Earth: the ECOLOGICAL REVOLUTION that began to accelerate visibly when World War I released a new wave of technology into the environment."[24] And it was precisely in emerging information technology that he believed a key was to be found. "The first dissident students at Berkeley," Youngblood proclaimed, echoing Fuller, "were born the year commercial television began."[25] What this meant, for both, was that this first "TV generation" were "native" to a new technological milieu—they participated in a global intermedia ecology operating within that videosphere and its interconnected "electronic information network." Out of this had seemingly emerged a new consciousness: "For the first time in history we can perceive Earth as a closed system, and thus can begin to use our life support system properly."[26] Central to the argument was not only that this global connectivity had led to a new "Whole Earth Consciousness" but that it had a political (or antipolitical) correlate: "youth of humanity all around our planet are intuitively revolting from all sovereignties and political ideologies," Youngblood went on to declare.[27] Fuller had made a similar remark when commenting on the student riots at UC Berkeley in the winter of 1964–65. The students, he argued, "were not interested in their state. They felt no loyalty to their nation. . . . Their idealism had lost its debilitating bias. They felt it to be immoral to be chauvinistic and patriotic. The young people were and are only interested in the whole world and in the welfare of all humanity."[28]

World Game sought precisely this "immediate dissolution of all sovereign nation state boundaries."[29] And it seemed that if the project could only harness the energy of discontented youth—whom he regarded as having an elective affinity with its ambition through their "Whole Earth Consciousness"—if it could actively provide a platform to develop this "intuition," then the revolution by design might succeed. And it could do so without state funding. Youngblood again:

> If every person who considered himself a citizen of the world, or a citizen of Woodstock Nation, would contribute one dollar, that would be a start. I'm trying to make you see that individuals and autonomous groups all around the world, apolitically, transcendental to politics, just using the technology that warmaking politicians gave us—we can achieve the downfall of the sovereign nation state technocracy and the birth of Whole Earth Unity.[30]

War as Extension of Politics

For Fuller what was at stake in the production of a universal humanity freed of the vicissitudes of time and place into a Dymaxion air-ocean world was thus not only a continuation of the Enlightenment narrative of man's liberation from the forces of nature through technology. When he argued that the "objective of the [World G]ame would be to explore for ways to make it possible for anybody and everybody in the human family to enjoy the total earth without any human interfering with any other human and without any human gaining advantage at the expense of another," he was arguing for the elimination of politics.[31] His 1967 manifesto for World Game, anthologized in *Utopia or Oblivion*, was framed through the antithesis of technology and politics. Referring at once to political theories, ideologies, organizations, and "concepts of political functions," all had to Fuller's mind become "obsolete" in the last decade. "All of them," he indicated, pointing to a notion of enmity, "were developed on the you-or-me basis."[32] (We might recall here Carl Schmitt's famous identification of the political in the friend / enemy distinction.)[33] That by raising the question of enmity Fuller was pointing to relations between politics and war was soon clarified. "If a team resorts to political pressures to accelerate their advantages," he wrote, "they are apt to be in trouble. When you get into politics you are very liable to get into war." To which he added, in a Clausewitzian turn of phrase, "War is the ultimate tool of politics."[34] War, von Clausewitz famously theorized, was a continuation of politics by other means; war was an "instrument of policy."[35]

Recall that for Fuller the war in Vietnam was an "experimental" war of ideologies, an attempt by the two superpowers—the United States and the Soviet Union—to wage war vicariously in a nuclear age without the threat of Armageddon.[36] What he called World War III served as the end point of a narrative traced from Thomas Malthus to Charles Darwin to Karl Marx, all of whom, to his mind, had remained trapped in the belief that there were not enough resources to go around, that only the fittest would survive. This belief, he argued, was the cause of an extreme mode of enmity premised on the total destruction, or absolute subjugation, of the enemy. It was, moreover, the reason that technological advancements had been developed for weaponry rather than livingry. World Game, although founded on a military game, was supposed to invert the very logic that led to absolute war. Fuller explained:

> In playing the game I propose that we set up a different system of games from that of Dr. John Von Neuman whose "Theory of Games" was always predicated upon one side losing 100 percent. His game theory is called "Drop Dead." In our World Game we propose to explore and test by assimilated adoption various schemes of

> "How to Make the World Work." To win the World Game everybody must be made physically successful. Everybody must win.[37]

Indeed, to resort to war, to "use the war-waging equipment with which all national political systems maintain their sovereign power," was to lose the game, to be disqualified.[38] The cold war arms race had, in any case, as Fuller acknowledged, altered the very terms of this equation. Relations between war and politics had to be rethought on account of new technologies. It was no longer that wars would end either through negotiating modified organizations of borders or territory or through one side fully dominating the other, since the situation had, in Fuller's words, "mushroomed to the point now where everyone is a loser in war with nuclear bombs."[39]

What, we might ask in retrospect, was wrong with this picture, with its noble ideals of a global humanity living free from want and illness, in a state of peace, equality, and harmony? Certainly the violent history of imperial conquests, characterized by the political domination and subjugation of other populations, and the exploitation of their resources, can be rightfully regarded as a legacy of the expansionist ideology of sovereign nation-states. And the increasing brutality of interstate warfare during the twentieth century, a condition fostered in part by technological "advancement," must haunt any assessment of relations between war and politics during this period, whether we are considering just or unjust wars. Furthermore, there was and remains a distinct urgency to the search for peaceful solutions to global conflicts, not least on account of the endgame implicit in the escalation of nuclear weapons. Finally, although far from completing any list of the violence accompanying warfare, we might point to the suspension of democratic processes and rights—products of the state's so-called monopoly on violence and its sovereign capacity to declare "states of emergency."[40] Indeed, it was precisely such political tools that had been deployed during this period not only in the name of peace or civility among states but as weapons turned against contestatory citizens (notably civil rights demonstrators and antiwar protestors). This was a period in American history thought by some to constitute nothing less than another civil war, one in which "law and order" were maintained through the increasing militarization of everyday life. On November 4, 1968, *The Nation* offered an editorial entitled "Clausewitz Updated," which addressed this transposition. "Everyone who can read is reminded a dozen times a year," the editorial began, "that war is the extension of policy by other means. This dictum," it continued, "needs to be domesticated: A nation that maintains warlike, aggressive counter-revolutionary policies abroad is a nation that will eventually pursue these same objectives at home. Thus domestic politics becomes the extension of war by other means."[41]

Significant here is not just the problematic loss of distinctions between interstate wars and civil or guerilla warfare, or even political struggles taking place within the civil domain of a democratic society, with respect to the state's use of violence. Also important is that Fuller was equally prepared to collapse these domains. Rather than questioning the state's use of violence against its own citizens, he chose to equate all political life with the prospect of war. Politics, whether internal to the state, in the international arena, or even in extra-parliamentary form, was simply reduced to the cause of warfare. Talking in San Francisco in 1967 he demonstrated this slippage:

> So what I want the young world to realize is that actually we're right in screaming that we ought not to have war, and that they'd like not to have war[;] the way you're going to get it is by design revolution where you do more with less. . . . There's a design revolution ahead instead of a bloody and political revolution. If you go out playing politics, it's going to end in guns because politics always consumes one side or the other.[42]

Furthermore, there was no scope in this picture for peace to emerge through democratic political or juridical processes (a key role of international institutions), but only on account of their abandonment. In *Utopia or Oblivion* Fuller invoked such a shift: "Finding their own political demonstrations for peace or their outright revolutions leading only toward further war," he recounted, "a few pioneers amongst the world students have joined up objectively with the heretofore only subjectively experienced do-more-with-less design-science revolution."[43] Here, perhaps, was Fuller's own peculiar updating of Clausewitz. Technology was not offered as a tool or extension for a politics seeking peaceful solutions but had rather come to stand in as its replacement. But contemporary technology, as Fuller himself knew all too well, was itself a key instrument of modern warfare.

Beyond Fuller's outright rejection of politics, which he saw as the source of war, was a more subtle elision operating in his notion of a global or universalized humanity. For this notion of humanity (and of a world without conflict) is perhaps less efficacious as a tool in the service of peace than it is an image of forced consensus, of a global police state free from any moment of dissensus.[44] We might return briefly to Schmitt, who noted that humanity was not "a political concept, and no political entity or society and no status corresponds to it."[45] If on the one hand this might have seemed to imply the overcoming of war (in the argument that "humanity as such cannot wage war because it has no enemy, at least not on this planet"), on the other hand, as Schmitt expounded, "the concept of humanity is an especially useful ideological instrument of imperialist expansion, and in its ethical-humanitarian form it is a specific vehicle of economic imperialism."[46] Informing the universalizing logic of Fuller's one air-ocean world, as in his infinitely repeatable dwelling units—from the 4D dwellings of the late 1920s to their post–World War II counterpart, geodesic domes—was a conception of a way of life that was unified, homogenous, and dictated by the forces of technological advancement. This would, of course, serve all too well in the service of capitalist "de-" and "reterritorialization."[47]

World Game was a logical extension of this universalizing epistemology to a postindustrial era of global networks and post-Fordist capitalism. And even at a historical moment characterized by growing discontent, what Herbert Marcuse famously termed the "Great Refusal,"[48] it did not seek politicized lines of flight or attempt to forge prospects for heterogeneity within milieu. Not only was there no pursuit of differentiated and alternative modes of life in this model of liberation, but there was no space for a political subject, or for political participation, or even for forming political communities. Difference would be nonexistent. Fuller's was a model of liberalism that, in its claims to overcoming hierarchies operating within the politics of sovereign nation-states, assumed that nothing would step in to fill the vacuum of power. Least of all did he foresee that passage toward a global form of sovereignty that Michael Hardt and Antonio Negri have theorized as Empire, one very much mediated through those invisible global networks of control to which he would cede humanity's fate.[49] And this condition, far from offering a paradigm of peace, would be premised on a perpetual state of war. As Hardt and Negri explained in turn, "To the extent that the sovereign authority of nation-states, even the most dominant nation-states, is declining and there is instead emerging a new supranational form of sovereignty, a global Empire, the conditions and nature of war and political violence are necessarily changing. War is becoming a general phenomenon, global and interminable."[50]

Domes

Youngblood's embrace of World Game was, of course, only one among many legacies of Fuller's thinking within the counterculture, prominent among which, as mentioned earlier, was the adoption of the geodesic dome. I want to look briefly at this phenomenon before concluding, for it offers further insight into the politics of (re)deploying military technology. In his Earth Day lecture at USC, Youngblood took care to distinguish World Game's utopian project of "inventing alternative futures" from the use of domes as a technology of dropping out. "You already have the beginning of these experimental lifestyle cities today in the drop-out communes and dome communities," he noted of the search for new urbanisms. Yet, as he continued, carefully noting the intended inversion, "Primarily they're filled with cultural freak-outs, casualties of the urban technology. So World Game will turn them into counter-cultural drop-ins."[51]

The dome-building movement emerged in May 1965 with the founding of Drop City near Trinidad, Colorado. The relation between Fuller and the Droppers was a sympathetic and seemingly mutually-reinforcing one. By their own account, Fuller had both inspired and encouraged their interest in deploying geodesic domes as alternative living spaces while lecturing in Boulder the previous April. This was precisely the moment when, following the protests at Berkeley at the end of 1964, Fuller began to recognize in the desires of this revolutionary movement a potential momentum for his own revolution by design. With America's youth culture on his mind, it was hardly surprising that Fuller took an interest in the use of his invention (until then deployed primarily as a weapon in the cold war) for such an alternative purpose.

Constructed through recycling waste—used two-by-fours, tarpaper, scavenged railroad ties, factory-reject plywood, bottle tops, junk cars—the Drop City domes arose literally, in the Droppers' own words, from the "garbage of America."[52] "Creative scrounging" was not, of course, part of Fuller's ambitions to harness and redirect the military-industrial apparatus to other ends. Yet he was happy to be identified with Drop City, awarding the commune the 1966 "Dymaxion Award for poetically economic structural accomplishments." The Droppers sent a communal note of thanks, reiterating their indebtedness: "We think it is the greatest thing that's ever happened to us. As we have told you, a speech you made several years ago was partly responsible for the conception of Drop City. Your award renews that inspiration."[53] John McHale, then executive director of the World Resources Inventory, revealed that there had been something of a case of mistaken identity. In a letter to the Droppers he wrote: "Professor Fuller has further suggested that you might consider the future possibilities of such 'shelter production' as your own local industry to help maintain the city."[54] The entrepreneurial nature of the proposal harbors a significant misrecognition of what was at stake in founding Drop City. The Droppers' exodus from extant economic structures was self-consciously political: it was an experiment in communal living outside the military-industrial "system" so central to Fuller's project. "People who give us money to make things," one Dropper noted, "prevent their money from being used to destroy things."[55] Survival would not be brought about through profitable work but through being dedicated to refusing to work within that system. As reported in *Avatar*, "we have attempted to create in Drop City a total living environment, outside the structure of society." Its "tribal unit" had "no formal structure, no written laws." "Each Dropper is free. Each does what he wants. No rules, no duties, no obligations. Anarchy. . . . Droppers are not asked to do anything."[56] When work was undertaken, for instance in the construction of the domes, it was not in the service of even a sustainability level profit but the expression of a positive desire. "We play at working. . . . We are based on the pleasure principle." The inhabitants of this "geodesic gypsy city" were far from Fuller's ideal nomadic subjects building lightweight shelters with which to meet the coming apocalypse.

Geodesic domes were not, however, the only aspect of Fuller's work that appealed to the Droppers: they also fell prey to his universalizing vision. "We hope to buy more land, build more Drop Cities all over the world, universe."[57] By August 1967 they could announce that "Already Drop City South is firmly established near Albuquerque, N.M.,"[58] soon followed by New Buffalo in Arroyo Hondo. "Soon domed cities will spread across the world, anywhere land is cheap—on the deserts, in the swamps, on mountains, tundras, ice caps," Peter Rabbit confidently proclaimed. "The tribes are moving, building completely free and open way-stations, each a warm and beautiful conscious environment. We are winning."[59] Such optimism would soon subside. Peter Rabbit, by his own confession, "brought down the hordes on Drop City" by inciting the mass media, catalyzing the commune's demise. News of Drop City and its photogenic geodesic domes spread quickly. Articles appeared in *Arts Magazine*, *Aspen*, *Architectural Forum*, the *New Yorker*, and elsewhere. The press might have been good for expanding Fuller's reputation within the counterculture, but it would not prove to be so for the integrity of the commune. By 1968 most of the original members had fled.

Fuller's patent for the geodesic dome and his arguments for its role in the stewardship of Spaceship Earth, would in turn be central to countercultural publications such as Steve Baer's *Dome Cookbook* of 1968—the first manual for the do-it-yourself dome builder—and Stewart Brand's *Whole Earth Catalog*, initiated the same year.[60] These were soon followed by Lloyd Kahn's two *Domebooks*, which appeared in 1970 and 1971. Kahn's vision was closer to Fuller's own and avowedly indebted to it. "Shelter production" was its driving force, his ambition being to make dome-building technology available to "young people" through establishing a system of open access to dome-building information. As recounted in a letter of April 1969, Kahn had been heading toward "a career of building big timbered homes on the California Coast" but after hearing Fuller talk at the Esalen Institute promptly quit his job, "the Design

Science Revolution foremost in mind."[61] He soon founded Pacific Domes and, along with Jay and Kathleen Baldwin and others, took part in the construction of ten domes at the Pacific High School in the Santa Cruz hills. Next came an "information booklet," initially titled "Domerise" but published in 1970 as *Domebook 1.* It included detailed diagrams, photographs, instructions, and helpful hints on techniques and materials for the would-be dome builder, typically cast through stories of dome construction. These served as "prototypes for future industrial production of low-cost housing."[62] Related to this was a proposal to develop therapeutic workshops, Kahn's avowed response to "the massive and growing student unrest": "Many young people these days either rebel or drift aimlessly because they cannot find meaningful outlets for their youthful energy," he wrote to Fuller. "With the proper skills, a student could after graduation go into the country, build a light dome, plant a garden." Finally was a proposal to gain corporate sponsorship for experimental dome-building communities. Identifying possible candidates as Alcoa and Union Carbide, he explained that the "basis for company approval" would be "that individuals will innovate where large organization cannot (as in your comparison of the Dymaxion car vs. Chrysler)."[63] In exchange for a place to live and test dome-building strategies, the companies would gain publicity and information on the use of its products.

That Fuller was pleased with this development was evident not only in his commissioning Baldwin to erect a pillow dome on Bear Island (like the one produced at Pacific High) but also in the response from Fuller's office in Carbondale. Writing to Kahn to propose a 5 percent royalty on net income from *Domebook* sales, Fuller's assistant, Dale Klaus, noted that "times are good and things are popping and you are, in some part, responsible for this."[64] The office would regularly direct inquiries to the office regarding geodesic domes to the *Domebooks.* Moreover, along with the *Whole Earth Catalog,* the *Domebook,* Klaus noted, had become a useful tool in sponsoring the World Game "attitude."[65]

Published in May 1971, *Domebook 2* was more than double the size of the first volume. But already Kahn, now acting as sole editor, was starting to be haunted by doubts: "There's a danger in the hype, overromanticizing domes," he warned readers.[66] A number of people had written questioning the integrity of the use of "space-age" materials, a "fall out from the space program" initially regarded as a very precondition for the emergence of the movement. If plastics, along with "Silicone caulks, polyurethane foam, clear ultra-violet resistant flexible vinyl," had initially been enthusiastically embraced, since trees were "critically needed for photosynthesis," the anxiety of deforestation was soon replaced both by the recognition that timber, unlike petroleum, was a renewable resource and that, beyond this, plastics were toxic. "The plasticizer molecules migrate, airborne into the dome," it was explained. "If you're thinking of building a plastic dome give it the acid test; get stoned, look hard, breathe deeply, feel the stuff."[67] Plastic, moreover, came to be recognized as part of a larger cycle of exploitation, one decidedly against the countercultural ethos. "Plastics are super products of western oil man sucking the earth dry of petroleum. Hucksters delude the American public into demanding products from oil: big hungry cars / plastic pin curlers / plastic wrapped food."[68]

By the time the third *Domebook* appeared, in 1973, subsumed within a larger volume entitled *Shelter*, those doubts had become something closer to a polemic against the construction of domes. Kahn's enthusiasm for Fuller's vision had, within the short span of a few years, come to a crushing halt. In a remarkable act of retrospection and atonement, *Shelter* offered a revised history. "We made an error in *Domebook 2* in stating that Buckminster Fuller was the inventor of the geodesic dome," he explained. "Fuller's contribution, rather than origination of the great circle principle, or its earliest structural utilization, is rather application of the word *geodesic* to this type of polyhedral building framework, and its popularization and commercialization in the United States."[69] Geodesics were now situated as the fifth and most recent type of dome, ancestors to which, according to this mythical tale, included the woven dome, an ur-dome of woven branches covered with thatch, leaves, or animal skins that for ideological or symbolic reasons had given rise to the wooden dome, the masonry dome, and then the Imperial Roman concrete dome. History was corrected not only in an introductory note but with an article, entitled "The Wonder of Jena," on Dr. Walter Bauersfeld's (unpatented) planetarium dome, built on the roof of the Carl Zeiss optical works in Jena, Germany, in 1922. Thirty years later, it was explained, Fuller had patented the "same subdivided icosahedron principle (in 1954)" and named the structures geodesic domes. To this was added "Smart but not Wise," Kahn's extended epistolary remarks on dome-building from 1971. Here he reiterated the critique of plastic offered by readers of the first *Domebook*: "In addition to the *practical* and *aesthetic* disadvantages . . . I've found in plastics there is the idea that one is dealing with Dow, and the oil industry—that is the people Nixon works for."[70]

Utopia or Oblivion?

When *Utopia or Oblivion* appeared in December 1969, it quickly enjoyed a wide circulation among a counterculture primed through Fuller's earlier publications and lectures, as well as through the *Whole Earth Catalog*. Bringing together writings and lectures dating from the previous half-decade, many continuing polemics from the 1920s, the volume was exemplary of Fuller's proselytizing for embracing the immanent development of technology, and it reiterated his long-standing claims that revolution might take place through unfettered technological advancement. Efficiency would simply trump politics in the fight for global survival. As indicated in the title, Fuller's message was cast not only through a utopian hope for humanity's prospects but through raising the specter of apocalypse. Youngblood's dedication to World Game was caught precisely between these very poles: on the one hand it was motivated by the sense of optimism Fuller inspired: "There are a lot of young people like me who were pretty uptight and didn't know what to do with their lives, who have now committed themselves to World Game," he explained. "It's really the only possible alternative for positive revolutionary action. . . . I can't think of anything more important to do with my time. I don't look back. And all I see ahead is hope."[71] But that hope remained bound to its dystopian counterpart. "The young lives of mid-century America find themselves perched on the fulcrum of a cosmic balancing act," he lamented, echoing Fuller, "with Utopia on the one hand and oblivion on the other."[72] This, he suggested, was "the last either / or in history."[73]

World Game seemed to offer a panacea, or at least a way to break this dialectic. Presented was a model of technology liberated from the "bias" of politics, a fluid world without nation-states, and the prospects of a globally "successful" humanity free from war. It is not difficult to see why this model of "how to make the world work" appealed to a generation raised on the threat of nuclear holocaust and radicalized by the war in Vietnam. But it seems less clear, in retrospect (the escalating disenchantment of the moment notwithstanding), why a generation politicized by those historical circumstances could have so quickly jettisoned the domain of the political, precisely the domain through which the civil rights struggles and antiwar movements had launched their contestations—jettisoned, moreover, at the expense of prospects for a positively cast, politically transformative notion of a better future.

Fuller's millenarianism also coursed through the culture of dome builders, themselves falling prey to an escalating rhetoric of insecurity. Working on the premise that only those who had found self-sufficient or autonomous ways of living would survive, dome building came to offer for some a means of testing not only new urban forms for a postrevolutionary society but strategies for surviving a massive environmental, and perhaps nuclear, catastrophe. That dome building was a product of the very system raising that apocalyptic threat had by 1971 become unacceptable even to those like Kahn who had done so much to promote it. His own trajectory would steer him away from what he called "white man technological prowess"—not, however, toward a form of political engagement with the military-industrial apparatus but, rather, headlong toward an increasingly paranoid survivalist mentality and a thoroughgoing rejection of technology (even to the prophecy of an inescapable return to handicraft).

The internal exodus of the counterculture from mainstream American society and politics had not, of course, come about through an intuitive embrace of technology without political conviction, as Fuller's narrative would have it. Ideals of justice forged within movements to create a "new America" did not emerge without contestation. But the search for a space for political engagement seems to have given way to a faith in technology that all but foreclosed prospects for forging politicized lines of flight. Faced with an environment increasingly taking on the logic of total war, in which weaponry was regularly retooled for livingry, responses to such accelerating technological development thus remained trapped within a dialectic of the uncritical embrace or paranoid rejection of technology. Without a model of political engagement, the embrace of Fuller's scenario of a "world-embracing and universe-ramifying evolution of industrialization"[74] yielded not only to the forces of technocracy but to the extant economic and political powers for whom such development served all too well, not as a means of equitably distributing world resources but in the pursuit of untrammeled profit. That is to say: far from effectively redirecting socioeconomic norms, such naturalization of technology as an evolutionary force implicitly supported, and continues to support, the very logic driving the expanding military-industrial complex and its capacity to produce ever more extensively networked forms of global power.

11 Fuller's Futures

Reinhold Martin

Often enough, one of the reasons given for revisiting the work of a figure such as R. Buckminster Fuller, whose historical significance is well established, is to draw lessons for the present or, indeed, for the future. Such is potentially the case here, with the celebration of an archive that promises, in effect, to keep Fuller's future-oriented memory alive for us in all of its complexity. But we also know that to remember is to forget, in the sense of the spatial phenomenon called fetishization, whereby fixation on certain memories necessarily screens out other, competing ones. Remembering can also bring on the temporal phenomenon called reification, whereby the fluid dynamics of history seem to be consolidated once and for all. In other words, the very act of archiving knowingly risks a Fuller frozen in space-time.

Fuller's Place

Having never made it into Siegfried Giedion's epic tale of modernism's historical voyage in *Space, Time, and Architecture* (1941), Fuller would seem somewhat immune to such a fate, which is normally reserved for modern architects proper as the traces they leave on history calcify over time. In the aftermath of world war Giedion himself would, in effect, write these heroes out of the story, in the "anonymous" dark history he called *Mechanization Takes Command* (1948). There, as if to make up for his absence in *Space, Time, and Architecture*, Fuller does make an appearance, as the least anonymous among the book's many supporting characters. Between its lines, Giedion's *Mechanization Takes Command* describes a state of affairs in which modern architects found themselves in a kind of freefall down the wind tunnel of progress. Still, these architects remained the very embodiment of a mythic—if now tragic—modernity, figured in Giedion's imagination as a forward-looking engineer. Just prior to the war, in a famous formulation, Walter Benjamin (himself a careful early reader of Giedion) would face a related figure backward to survey the debris-field of "progress" piling up in his wake and call him an angel: the angel of history.[1] In the decades following that calamity—the scope of which Benjamin could only have guessed at—Fuller would, in turn, turn history's angel back around to face forward again, calling him (or her) an astronaut, the pilot of Spaceship Earth.

To the degree that Earth's astronaut was also a cosmonaut, the future posited by Fuller in his *Operating Manual for Spaceship Earth* (1969) and related works can seem decidedly postpolitical when seen against the backdrop of the cold war, in which Fuller himself was an active participant.[2] And, indeed, it does share a certain family resemblance with such ventures as Daniel Bell's "end of ideology" and the coming of a "post-industrial society," despite Fuller's protestations at being labeled a technocrat.[3] Still, for reasons that I hope will become evident, I would like to approach the question of the future as it appears in Fuller's later work from a slightly different angle. That angle is generously afforded by an otherwise preposterous question: Was Buckminster Fuller a postmodernist?

This question, at minimum, is authorized by a quirk of chronology, since Fuller's prodigious career spanned the time periods generally associated, in architecture, with high modernism (marked, say, by his Dymaxion house of 1927) and then with postmodernism (said by Charles Jencks to have begun on July 15, 1972, at 3:32 p.m., with the demolition of Minoru Yamasaki's Pruitt-Igoe housing complex).[4] Technically, this makes the post–World War II Fuller of the geodesic domes a "late modernist" in Jencks's eyes. The stylistic marker found in his work (one of many possible "late modernist" traits, according to Jencks) is an extreme repetition—"the sizzle of incessant space frames" shared with Philip Johnson and Cesar Pelli among others, and synchronic with both the "wet look" of James Stirling at Olivetti and the "slick skins" of Norman Foster, again among many others.[5]

Following in the footsteps of Jencks, Fredric Jameson, a signal theorist of cultural postmodernism (or what he calls the "cultural logic of late capitalism") has described such "late moderns," in architecture, literature, and elsewhere, as transitional figures. For Jameson the category of the "late modern" would initially contain "the last survivals of a properly modernist view of art and the world after the great political and economic break of the Depression."[6] Later, it would come to mean the international artistic ideology of the cold war, epitomized by the postutopian "American" formalism of Clement Greenberg, which sought definitively to separate art from politics and, most of all, from a vulgarized culture industry.[7] All of this seems a far cry from Fuller, whose profile as a self-proclaimed artist, engineer, ecologist, and "comprehensive designer" hardly fulfills the imperatives of a Greenbergian autonomy of art, which Jameson sees as an immediate precursor to a fully postutopian—that is, a posthistorical, postpolitical—postmodernism. On the contrary, Fuller's synthetic, "synergetic" efforts to map the entire "world-around" system in a manner adequate to its pilot's navigational and cartographic requirements would seem to place him on the other side of a fragmented, postmodern division of labor in which everyone merely does his or her own job without access to what Jameson calls cognitive maps.

Jameson extrapolates his analytic of cognitive mapping from Kevin Lynch's *Image of the City* (1960), a work devoted to analyzing different forms of what Lynch calls "imageability" at the urban scale. In this context a cognitive map is a mode of representation that allows the inhabitant to grasp the totality of the city and discern his or her place within it. Jameson extends Lynch's model to apply to the world system of late capitalism in general—an intensely disorienting space lacking familiar guideposts and thus requiring new mental "maps" to grasp its scope and its momentum.[8] A similar cartographic impulse, arising out of an effort to deal with an increasingly complex and abstract global environment, might well be attributed to Fuller—for whom such maps would be indispensable to any effort to ascertain the direction in which Spaceship Earth is heading and to correct its course.

However, to the degree that Fuller also replaces the starkly ambivalent choice offered by Le Corbusier—"Architecture or Revolution"—with an even starker if far less ambivalent one—"Utopia or Oblivion"—his overall project would seem to signal something like the exaggerated persistence of a utopian "high modernism" (born, like Fuller, in the late nineteenth century) *within* the very fabric of that disorienting hall of mirrors called postmodernism. From both a historical and an epistemological perspective, this impulse corresponds most closely to what Jameson views as the symptomatic nature of 1970s science fiction. When not subjected to an unsentimental shredding in the hands of a Philip K. Dick or a J. G. Ballard, the utopianism characteristic of much science fiction at the time was frequently modeled on the secessionist politics of the 1960s counterculture. Two examples of this were Ursula K. Le Guin's *The Dispossessed* (1974) and Ernest Callenbach's *Ecotopia* (1975). Both novels described the production of parallel mirror worlds in which were reflected negatively the excesses of consumer society, through a Fulleresque harnessing of postindustrial knowledge (to different degrees) for the equitable, sustainable redistribution of resources. In Le Guin's *The Dispossessed*, the mirror world took the form of the moon Annares, a barren anarchist satellite orbiting the Earth-like Urras. Callenbach's *Ecotopia* names a breakaway republic consisting of what was once Northern California, Oregon, and Washington, now devoted to a life lived in systematic, unrelenting harmony with nature.

Jameson sees the value of such visions to lie not in the dubious and often ambiguous alternatives to the postmodern status quo they offer but in their ability to sponsor a negative dialectic in which their opposite number's ideological trappings are fully exposed.[9] In these negative mirror worlds, the motives and protocols observed by those at the helm of Spaceship Earth below or beyond are revealed to be grossly unjust. So, too, does the present get reflected in Fuller's own future worlds but in a slightly different way. By the 1970s, that is, Fuller was given not so much to projecting ideal futures as to playing games with them.

World Games and Their Rules

For example, in Fuller's World Game (begun in 1965), we see a technical, cartographic instance of what Jean-François Lyotard, another central theorist of postmodernism, calls "language games." For Lyotard (following Ludwig Wittgenstein) a language game is an experimental game played with linguistic codes and protocols at various levels. To speak or write either denotatively or prescriptively is to play such a game, and to violate its codes (or to tell a different kind of story) is to innovate.[10] From this perspective we might notice that the World Game turns technocratic positivism into a playful experiment with a series of different narrative scenarios, where the more radically a given scenario rearranges global technoeconomic assumptions, the more likely it is to offer an alternative to the status quo. However, what differentiates it from Lyotard's effort to replace overarching, modernist master narratives with a plurality of competing *pétits-récits* (small stories) is that the World Game is premised on the abilities of its players to grasp—and thereby to manage and to direct—the totality of the dynamic world system with the help of maps. In the balances of trade and other quantitative assessments measured and rearranged in the World Game, what is being contested is not this or that micronarrative corresponding to what Lyotard calls (after René Thom) an "island of determinism" in the heaving, directionless postmodern sea.[11] Rather, the World Game posits a set of competing master narratives that tell the story of the global future as such.

In that regard Fuller's futures surely reflect the totalizing futurology of the general systems theory that lay behind them, which Lyotard would condemn at the end of the 1970s.[12] But the World Game is a game played with the possibility of reimagining the future as such, as well as a game of possible futures. Therefore, Fuller's project is not reducible to an enterprise that, as Lyotard warns about systems theory, merely proposes "a 'pure' alternative to the system" that "would end up resembling the system it was meant to replace."[13] Instead, Fuller's project reorients the "system" from within by playing games with the very *idea* of a graspable, collective future.

As Fuller designed it, in the World Game individuals or teams "would each develop their own theory of how to make the total world work successfully for all of humanity."[14] This would be done on an electronic version of his "sky-ocean" Dymaxion map originally designed (but unrealized) as a collapsible "geosphere" within the geodesic dome he built for the United States Information Agency for Expo 67 in Montreal. The format was modeled on the war games played by cold warriors, with the notable difference that the Manichean "Drop Dead," zero-sum premises of the former (based on mathematician and computer scientist John von Neumann's game theory) were to be replaced by the distinctly Fulleresque formulation: "Everybody must win."[15] This was utopian, to be sure, but with a certain tautological precision. Since if the objective of the game was to devise a redistribution of resources in which everybody wins, it was nevertheless impossible to win the World Game, not because this "ideal" scenario was permanently out of reach but because its availability was premised on an agonistics of knowledge (playing the game to win by devising the "correct" scenario) that, from the beginning, cancelled the synergetic cooperation necessary for all players to win.

Another way of saying this is that the World Game was played on two contradictory levels at once—one intrinsic, another extrinsic. Intrinsically, it was a kind of postmodern language game, in which no one scenario had an a priori metaphysical or empirical claim over any other. Nor did it assume any power differential among the players (in other words, a politics of knowledge), in which sense it was "postpolitical." Extrinsically, on the other hand, it remained thoroughly modernist, in the sense that it posited a space—mapped and modeled by the geodesic dome itself—in which something more than a temporary consensus could be reached, once the giant computer had, with the help of its human "players," played out all the possible scenarios. At this extrinsic or external level, the World Game was a modernist game of optimization at the scale of the world system itself, rather than a postmodernist game of perpetual, competitive innovation.

This would also mean that, extrinsically, the World Game was not as postpolitical or postideological as Fuller liked to claim. On the contrary, it entailed a displacement of politics to the level of cartography. It was a roadmap to a utopian future but one in which the political question was, in part, that of who was in charge of the cognitive maps. For Fuller himself this was a nonquestion, comparable to asking who was flying the many airplanes in which he circled the globe. The ultimate arbiter in the World Game would be the mainframe computer rather than a political entity. As Fuller put it, "What I proposed was based on my observation that world people had become extraordinarily confident in the fundamental reliability of the computer and its electronically controlled processes," a state of affairs verified by "the equanimity with which world-around air jet travelers now commit their lives to the computer's reliability" as they come in for a night landing.[16]

This presupposed, of course, that the destination toward which Spaceship Earth ought to be headed was preprogrammed or, to put it another way, that the utopian future could be represented transparently and thereby optimized. In contrast to the high modernist utopias of Le Corbusier, for example, which were represented in panoramic aerial views and integrated master plans, Fuller's futures were represented discursively and probabilistically, in charts, graphs, and statistics describing world-historical "trending" (his term). Still, it was assumed that these documents, famously archived at his "headquarters" at Southern Illinois University and now held at Stanford, were themselves uncontestable, in the sense that they represented objective trends rather than an ideological project. At one level this was nothing more than raw positivism. But at another it was a wager. The stakes of the World Game did not really lie in the question of whether the statistics it offered were scientifically verifiable and therefore constituted a solid foundation on which an optimal future could be constructed, whether agonistically or consensually. Instead, the stakes lay most profoundly in the conversion of modernist utopias of form (Le Corbusier) into postmodernist utopias of risk (Fuller).

In 1986 the sociologist Ulrich Beck coined the term "risk society" to describe what he considered to be a second order or "reflexive" modernity organized around the social relations of statistically calibrated risk. Here we can think of Fuller's formula, "Utopia or Oblivion," as designating the horizon of a regime in which, as Beck puts it, "the risks of civilization today *escape perception* and are localized in the sphere of *physical and chemical formulas*."[17] As with Fuller, many of Beck's examples apply to environmental risks measured in charts, graphs, and mathematical formulas—rather than perceived directly—which serve as a kind of emblematic instance of the overall situation. That Beck used the phrase *reflexive modernity* rather than *postmodernism* to describe this is less important than the correspondences of his thesis with then-evolving hypotheses regarding cultural postmodernism. For example, as with Jameson's reflections on the structural nature of speculative finance capital and risk-reward calculations under postmodernity, under Beck's reflexive modernity the future appears as a set of probabilities:

> The center of risk-consciousness lies not in the present, but *in the future*. In the risk society, the past loses the power to determine the present. Its place is taken by the future, thus, something non-existent, invented, fictive, as the "cause" of current

experiences and action. We become active today in order to prevent, alleviate, or take precautions against the problems and crises of tomorrow and the day after tomorrow—or not to do so.[18]

Thus the projection of necessarily fictional future scenarios is constitutive of, rather than opposed to, the present. In other words, in risk society, as in postmodernism, the science-fiction future is a feedback loop.

In this light Fuller's defense of large-scale planning at a moment when master plans and the master narratives that authorized them were already coming under attack was not a modernist throwback. In his *Operating Manual*, as elsewhere, Fuller advocates thinking and planning at the scale of the universe itself:

> We are faced with an entirely new relationship to the universe. We are going to have to spread our wings of intellect and fly or perish; that is, we must dare immediately to fly by the generalized principles governing universe and not by the ground rules governing yesterday's superstitious and erroneously conditioned reflexes. And as we attempt comprehensive thinking we immediately begin to reemploy our innate drive for comprehensive understanding.
>
> The architects and planners, though rated as specialists, have a little wider focus than do the other professions. . . . At least the planners are allowed to look at *all* of Philadelphia and not just to peek through a hole at one house or through one door at one room in that house. So I think it's appropriate that we assume the role of planners and begin to do the largest scale comprehensive thinking of which we are capable.[19]

Fuller's version of this comprehensive perspective is given in general systems theory, which describes the "general system" in terms of interacting variables subject to parametric quantification. Here, we can think of the resources mapped and tracked in the World Game: population, energy, shipping lanes, railroads, transoceanic cables, airways, airports, satellites, television receivers, universities, literacy, vegetables, bread, motor vehicles, copper, earthquakes, electrical networks, and so on. If this list already seems impossibly incomplete or composed mainly of incommensurables, the issue is—again—less the capacity actually to account for, map, and accurately predict future trends than the capacity to project alternative futures based on the "fictions" (Beck) narrated by these maps. Further, the Fulleresque utopias conjured to avert the oblivion otherwise risked by Spaceship Earth—domes over Manhattan, Tetra-Cities, and floating geodesic "clouds," to name a few—are to be appreciated not so much for their capacity to organize the *future* as to organize the *present*, that is, where Fuller's futures are played out—in a postmodern present of which he and his work are inextricably a part but from which they are always busy planning an escape. Hardly escapist, however, Fuller's utopian fictions, both mapped and designed, confront us squarely with our own place within the placeless universe called postmodernity. Thus, if theorists like Jameson would call for cognitive maps to help navigate our way out of its hall of mirrors, Fuller always had such maps ready at hand, in the World Game and elsewhere.

And so we return to the aporia of the World Game—the simultaneous staging of multiple perspectives and multiple futures intrinsically, which are to be resolved into a single perspective (that of the spaceship's astro-cosmonaut), extrinsically. The purpose is to give direction, to steer the spaceship toward better prospects. The result, however, is to restage, again and again, what Jameson calls the postmodernist "crisis of the future"—the apparent impossibility of imagining a way out. Despite the faith exhibited by Fuller and his fellow astro-cosmonauts in the computers guiding the "world-around" machines (or *systems*) in which we metaphorically and literally fly to this day, the question remains: Where are we going?

In returning to this question—which is situated somewhere between Benjamin's angel of history and Fuller's astro-cosmonaut—we are not done playing the "game" of the World Game, whether it is conceived according to a computerized calculus of universal justice or a postmodern "just gaming" (Lyotard).[20] There is another aporia written into Fuller's futures. This is the aporia of the "we" in the question just posed, the players of the World Game and especially, the figure of the "world."

Who is it that plays this game, and with what are they playing? In practice, the players were mainly students, who were nevertheless playing with the very conditions of possibility for the "world" itself—that is, a form of collectivity that could emerge out of postmodern, "world-around" dispersal and deterritorialization. Fuller's futures were organized around the risk of "oblivion," in the face of which they sought a utopian coming-together of something like a world-system, a system of systems mapped cognitively in Dymaxion triangles and geodesic spheres, that was analogized to the "universe" of Fuller's special, quirky brand of systems theory—a literally *universal* world-system.

Ironically, the name of this world-system for postmodern thinkers, like Jameson, still willing to think totalities is late capitalism: a consumerist acceleration of postindustrial, informatic exchange and flexible accumulation also displaced onto the level of culture. Today it is more commonly named *globalization.* In architecture perhaps the most compelling figure for this "new machine" (as Jameson calls it)—literally, in Fuller's case, the globe—of which "we" require cognitive maps in order to exit, was given in 1984 in Jameson's reading of John Portman's 1977 atrium-equipped Bonaventure Hotel in Los Angeles. According to Jameson, that building "aspires to being a total space, a complete world, a kind of miniature city; to this new total space, meanwhile, corresponds a new collective practice, a new mode in which individuals move and congregate, something like the practice of a new and historically original kind of hypercrowd."[21] Let us insist, then, that Fuller's geodesic domes, and especially his dome for Expo 67, with its proposed geoscope / World Game inside, represent another, more technical iteration of such a space, in the form of an atrium enclosed in a sizzling space frame. Remember, too, that Jameson, immediately after describing the multiple disorientations of Portman's building, declares himself "anxious that Portman's space not be perceived as exceptional or seemingly marginalized and leisure-specialized on the order of Disneyland"[22]—or, we could add, on the order of world expositions like Expo 67 or like Disney*world*, with Epcot Center as its geodesic centerpiece.

But more than merely mapping the globe and thereby constructing something like a global subject made up of World Game players, Fuller's geodesic domes traversed it—often as emblems and instruments of world trade and / or geopolitics. There were, for example, the postcolonial domes erected in 1957 for the United States Department of Commerce conference in Kabul, and the Calico Company geodesic dome pavilion in Bombay of 1958 (fig. 11.1). And,

Figure 11.1
U.S. Department of Commerce dome in Kabul, Afghanistan, erected in 1965.
Source: Special Collections, Stanford University Libraries.

Figure 11.2
Proposed geoscope, a two-hundred-foot-diameter miniature Earth that would be suspended one hundred feet above the ground and fitted with miniature lights to provide a visual representation of real-time world data. Source: Special Collections, Stanford University Libraries.

most dramatically, there was the U.S. pavilion at the American National Exhibition of 1959 in Moscow, in which was projected the sweetly propagandistic multiscreen "Glimpses of the USA" by Charles and Ray Eames. All of these were precursors to the Expo 67 dome with its projected geoscope (fig. 11.2). So the geodesic dome did not simply *represent* the world. Like Fuller, it traveled across its surfaces, tracing numerous interrelated paths along the great circles that girdled it. Thus, in contrast to the many earthbound, monocular architectural domes—Eastern and Western—that preceded it over many centuries, the geodesic dome also combined multiple *imagines mundi* with a logistics of reproduction and distribution (Fuller patented various designs) that effectively played the game of the world by literally drawing it out—encircling it many times, with Fuller flying alongside—in real space and real time.

Spaceship Earth, the "world" of the World Game and of the geodesic domes, was therefore a replica of the postmodernist "new machine." It, too, was a mirror world, reflected thousands of times in the thousands of geodesics that circled its surfaces during the latter part of Fuller's illustrious career. Its definitive monument was, perhaps, the "Fly's Eye Dome" of 1965—a dome made up of dozens of other, smaller dome-shaped units that refracted the image of the globe at finer scales (fig. 11.3). Here, the aporetic game of inside and outside played by the World Game was reversed. Instead of an internal multiplicity shaped externally by the machine's numerous feedback loops into a single "winning" scenario, and instead of a unitary spaceship piloted by a unitary astro-cosmonaut, suddenly and jarringly, an external totality—one world, one future—is fed back into the machine itself. And as in a fly's eye, the machine reproduces that totality across its reflective surface, to yield many worlds and many futures played out inconclusively by many spaceships with many pilots with many eyes, side by side.

These were Fuller's futures: a modernist utopia of postmodern replication. On the one hand, this utopia consisted of multiple, competing futures played out inside a spaceship-dome by multiple players. On the other hand, it con-

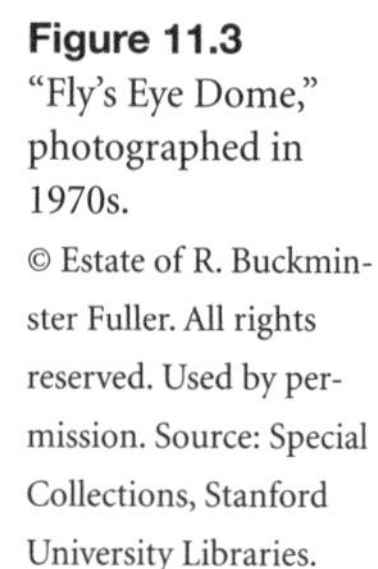

Figure 11.3
"Fly's Eye Dome," photographed in 1970s.
 Source: Special Collections, Stanford University Libraries.

sisted of a single, optimal future, a path through outer space (or history) taken by the spaceship-dome and plotted by the players inside. It would be a mistake to defend or dismiss such a project on the grounds of its success or failure in synthesizing these two levels, in deriving consensus from dissensus, or prescriptions from descriptions. As with all the geodesics, the most challenging design problem remained where to put the windows, not to mention the door. The question is not just where Spaceship Earth is heading nor even who is at the helm. The question is how to escape the self-reflexive, self-contained regime of risk management, the hall of mirrors in which the entire game is played. After all, what is ultimately at risk here is the possibility of imagining the future as a way out—a way out of the globe, the spaceship, the hotel atrium, the wind-tunnel of progress, and the geodesic dome itself, with its sizzling space frames and its impossible World Games.

Reference Matter

Notes

Introduction

1. R. Buckminster Fuller, *Guinea Pig B: The 56 Year Experiment*, repr. ed. (Clayton, CA: Critical Path Publishing, 2004).

2. The size of the archive is conservatively estimated at fourteen hundred linear feet of paper-based materials, manuscripts, and recordings.

3. Reinhold Martin, "Forget Fuller?" *ANY* magazine, no. 17 (Jan. 1997): 15–55, 15.

4. Barry M. Katz, in his chapter in this volume, points out that Fuller "cultivated his personal narrative to what can only be described as mythic proportions." Maria Gough recounts in her piece that by the 1950s "the proper name *Fuller* referred as much to a corporation as to an individual."

5. For example, in the 1930s Frank Lloyd Wright proposed creating affordable, single-story homes for middle-income Americans. His so-called *Usonian* houses would be used to populate small suburban or semirural planned communities.

6. Joachim Krausse, in this volume, references William Kuhns's remark that "Richard Buckminster Fuller is a nineteenth-century inventor with twenty-first-century ideas. The fact that he lives in the twentieth century seems a dual anachronism." See William Kuhns, *The Post-Industrial Prophets* (New York: Weybright and Talley, 1971), 220.

7. R. Buckminster Fuller, in *What I Have Learned* (New York: Simon and Schuster, 1968), 73.

Chapter 1

1. Lloyd Steven Sieden, *Buckminster Fuller's Universe: An Appreciation* (New York: Plenum, 1989), 70.

2. The modifier *Dymaxion* was probably added later since the term was coined by a public relations professional at Marshall Field's department store when one of Fuller's model homes was exhibited there in 1929. *Dymaxion* is a synthesis of the words *dynamic*, *maximum*, and *tension*.

3. Sieden, *Buckminster Fuller's Universe*, 70.

4. R. Buckminster Fuller, citing Oregon Lecture #9, July 12, 1962, in R. Buckminster Fuller, *Synergetics Dictionary: The Mind of Buckminster Fuller* (New York: Garland, 1986), 1:263.

5. Fuller, *Synergetics Dictionary*, 1:647.

6. Bonnie Goldstein DeVarco, "Life, Facts and Artifacts," available online at www.thirteen.org/bucky/devarco.html (accessed Sep. 29, 2008).

7. R. Buckminster Fuller, *The R. Buckminster Fuller Archives*, to Morris Library, Southern Illinois University, May 1, 1965, R. Buckminster Fuller Collection, M1090:2:236, Stanford University Libraries (hereafter cited as Fuller MSS).

8. R. Buckminster Fuller, *Bucky* (1966), Fuller MSS, M1090:1:2.

9. Ibid.

10. Isamu Noguchi, Marshall McLuhan, Shoji Sadao, Constantine Doxiadis, Henry Dreyfuss, and Charles Eames are just a few of the people with whom Fuller corresponded.

11. Sieden, *Buckminster Fuller's Universe*, 2.

12. There is some confusion about the exact date that the so-called Dymaxion Chronofile was started. In a July 1995 taped interview, the then-archivist of the Fuller Archive, Bonnie Goldstein DeVarco, suggested that the Chronofile was started in 1927, and all volumes prior to that were retroactive compilations. Fuller's 1965 document clearly states 1917 as the date of conception of the Chronofile, although it would not be known by that name until later. (The name "Dymaxion Chronofile" was likely given during or after 1929, when the word *Dymaxion* was coined for an exhibition of Fuller's 4D house at Marshall Field's department store in Chicago.) In either case, it is clear that Fuller had been saving significant amounts of correspondence, clippings, and other materials since even before 1917.

13. The papers include but are not limited to correspondence, telegrams, receipts, notes, telephone messages, and newspaper clippings.

14. Sieden, *Buckminster Fuller's Universe*, 7.

15. The Stockade Corporation manufactured lightweight building blocks made of compressed fibers and had a system for quickly erecting structures using the blocks. The company failed to make a profit and was bought out by Celotex Corporation in 1926; in the reorganization Fuller lost his position as president and ultimately was pushed out of the company.

16. Fuller, *Bucky*, n.p.

17. Ibid.

18. See Ernest Becker, *The Denial of Death* (New York: Free Press, 1973), 3.

19. Hal Aigner, "World Game," *Mother Earth Magazine* 6 (Nov./Dec. 1970), available online at www.motherearthnews.com/Do-It-Yourself/1970-11-01/World-Game.aspx?page=1 (accessed Dec. 8, 2008). In the same article the author explains Gabel's qualification: "Gabel, a 24-year-old design science student at Southern Illinois University, was later to amend his remarks, noting that 'proof' is an ambiguous term—some people won't have anything proven to them no matter how much evidence is presented—and much of his reference material was in files and would take a couple of months to assemble."

20. Robert W. Marks, "Bucky Fuller's Dymaxion World," *Science Illustrated* (reprint, pages unnumbered), in Fuller MSS, M1090:3:1.

21. R. Buckminster Fuller, "Everything I Know" lecture series (Jan. 1975). Transcript available online at www.bfi.org/our_programs/who_is_buckminster_fuller/online_resources/session_06 (accessed Dec. 8, 2008).

22. Becker, *The Denial of Death*, 3.

23. Ibid., 15.

24. Fuller typically charged from $2,000 to $5,000 as a speaking fee per engage-

ment. Particularly during the later part of his career, he was generally unwilling to negotiate his fee. See also Fuller MSS, M1090:3:1.

25. Roald Dahl, *Going Solo* (New York: Farrar, Straus and Giroux, 1986), n.p (front matter).

26. R. Buckminster Fuller to Ralph McCoy, Oct. 3, 1960, Fuller MSS, M1090:2:136.

27. Ralph McCoy to R. Buckminster Fuller, May 5, 1965, Fuller MSS, M1090:2:136.

28. Gerald Cross, *Annual Progress Report, R. B. Fuller Archives*, May 1965, Fuller MSS, M1090:2:136.

29. Southern Illinois University Special Research Project Allotment Fiscal Year 1964–65. Fuller MSS, M1090:7:19.

30. Marleen Trout, "Remembering SIU's Legend," *Daily Egyptian* 86, no. 111 (2001).

31. Thomas Turner to R. Buckminster Fuller, July 21, 1971, Fuller MSS, M1090:2:235.

32. Thomas Turner to Herman Wolf, Aug. 23, 1971, Fuller MSS, M1090:2:235.

33. R. Buckminster Fuller, written statement, 1972, Fuller MSS, M1090:2:236.

34. R. Buckminster Fuller to Ralph McCoy, Sep. 8, 1972, Fuller MSS, M1090:2:236.

35. Fuller to McCoy, Feb. 11, 1971, Fuller MSS, M1090:2:216.

36. Ed Applewhite to Gerard Dickler, April 5, 1971, Fuller MSS, M1090:2:216.

37. Fuller, written statement, 1972, Fuller MSS, M1090:2:236.

38. R. Buckminster Fuller to Charles Sachs, Oct. 5, 1979, Fuller MSS, M1090:2:517.

39. Fuller to Sachs, Jan. 14, 1980, Fuller MSS, M1090:2:517.

Chapter 2

1. R. Buckminster Fuller, "Later Development of My Thought," in *Ideas and Integrities: A Spontaneous Autobiographical Disclosure*, ed. Robert Marks (Englewood Cliffs, NJ: Prentice-Hall, 1963), 45.

2. Ibid.

3. "My work was initiated in 1927 and was designed to become effective in 1952" ("Design Strategy," in R. Buckminster Fuller, *Utopia or Oblivion: The Prospects for Humanity* [New York: Bantam, 1969], 318). On his supposed vow of silence see Fuller, "Later Development of My Thought," 47.

4. These short passages are culled from R. Buckminster Fuller, *Critical Path* (New York: St. Martin's, 1981), 124; from his notes archived in the R. Buckminster Fuller Collection, M1090:8:89:6 and M1090:8:89:8, Stanford University Libraries (hereafter cited as Fuller MSS); and from his essay "The Music of the New Life" (1964), published in Fuller, *Utopia or Oblivion*, 42, respectively.

5. Fuller MSS, M1090:8:89:6.

6. In the midst of the crisis year of 1927, "there came to my assistance the logic and fundamental sense, inherent from my long association and natural love of the sea and its boats" (R. Buckminster Fuller to Vincent Astor, Aug. 28, 1928, Fuller MSS, M1090:2:20:35).

7. The packing house—Armour & Co.—also gave this blue-blooded New Englander his first exposure to the "sacrilegious" jabbering of "jews & polaks," but this indignity does not appear among the "ideas and integrities" that marked his later recollections (R. Buckminster Fuller to Anne Hewlett Fuller, Jan. 12, 1916, Fuller MSS, M1090:2:45:2).

8. R. Buckminster Fuller, "Influences on My Work" (1955), in Buckminster Fuller, *Ideas and Integrities: A Spontaneous Autobiographic Disclosure* (New York: Macmillan, 1963), 14.

9. Fuller to George N. Buffington, Aug. 31, 1928, Fuller MSS, M1090:2:20:35.

10. Fuller to Anne Hewlett Fuller, March 17, April 1, May 3, 1927.

11. The preceding passages are drawn from, respectively, Fuller, "Later Development of My Thought," 61; R. Buckminster Fuller, *Grunch of Giants* (New York: St. Martin's, 1983), xi; Fuller MSS, M1090:89:8 (1973); Fuller, "Later Development of My Thought," 59; and "Guinea Pig B," Introduction to *Inventions: The Patented Works of R. Buckminster Fuller* (New York: St. Martin's, 1983), xiii.

12. A model of his first Dymaxion house was displayed at the Modern Decorative Arts Exhibition, sponsored by the Marshall Field's department store in April 1928. The announcement of the exhibit, "which introduces scientific application of space-time principles and harmonics into modern building construction," concludes: "Mr. Fuller cordially invites you to hear his short lecture, which will be given daily, each hour from 12 PM to 5 PM." It is possible that Fuller later conflated his profound depression of 1922 with the events of 1927.

13. Fuller to Buffington, Aug. 31, 1928, Fuller MSS, M1090:2:20:35. The fact that Fuller offers dramatically different accounts of this defining moment casts further doubt on the reliability of the mature Fuller as a source for his younger self. See Robert Snyder, *Buckminster Fuller: An Auto-biographical Monologue Scenario* (New York: St. Martin's, 1980), 35.

14. Fuller MSS, M1090:2:20:34, April 11, 1928.

15. Fuller MSS, M1090:2:20:35, July 16, 1928.

16. The manuscript went through three phases: first came an initial, seventy-three-page draft, entitled "Lightful Houses," which was completed by January or February of 1928. *4D* was produced in April and was the version that Fuller distributed in a privately printed edition of two hundred; a final version, later that year, incorporates correspondence and feedback.

17. In July Anne finally sublet their New York apartment and joined Bucky in Chicago, where, most obligingly, she began to keep a diary. In an entry for January 30, 1928, she noted that "RBF read Le Corbusier until very late at night" (Fuller MSS, M1090:2:15:30). *Vers une architecture* had just been translated into English and seems a likely inspiration for Fuller's thinking. Further links to Le Corbusier and other European modernists are explored by Michael John Gorman, *Buckminster Fuller: Designing for Mobility* (New York: Rizzoli, 2005), chap. 1.

18. Fuller MSS, M1090:2:15:30.

19. The typescript of *4D* was produced in great haste and is full of misspellings, grammatical curiosities, and terminological inconsistencies. This applies to the title itself, which at times is rendered *4D* and at times *4-D*, often on the same page. The observation by Marshall Field's branding expert, Waldo Warren, that this sounded more like an apartment number than a proposal for the restructuring of civilization led in the following year to the term *Dymaxion.*

20. See R. Buckminster Fuller, *Your Private Sky: R. Buckminster Fuller, the Art of Design Science*, ed. Joachim Krausse and Claude Lichtenstein (Baden: Lars Müller,

1999), referenced hereafter as vol. 1 of *YPS*; R. Buckminster Fuller, *Your Private Sky: R. Buckminster Fuller, Discourse*, ed. Joachim Krausse and Claude Lichtenstein (Baden: Lars Müller, 2001), referenced hereafter as vol. 2 of *YPS*; J. Baldwin, *Bucky Works: Buckminster Fuller's Ideas for Today* (New York: Wiley, 1966); Robert Marks, *The Dymaxion World of Buckminster Fuller* (New York: Reinhold, 1960); Martin Pawley, *Buckminster Fuller* (London: Grafton, 1992); Lloyd Steven Sieden, *Buckminster Fuller's Universe: An Appreciation* (New York: Perseus, 2000).

21. Fuller to Buffington, Aug. 31, 1928, Fuller MSS, M1090:2:20:35.

22. Fuller MSS, M1090:2:20:35 (1928): 4D, p. 2.

23. Fuller's comment about "assininic [*sic*] aesthetics" comes from a letter to Victor Astor, Aug. 8, 1928 (Fuller MSS, M1090:2:20:35); the "rhythmical contemplation of life" is from *4D Time Lock*, 35. In a letter to Bucky's bewildered brother, Wolcott, who, despite the sincerest filial sympathies was unable to make "head or tail out of the book," Anne explained that "we both feel it will be the entire solution of so many things. . . . The effects on land values, politics, almost all the manners and customs of today will be tremendous" (Anne Hewlett Fuller to Wolcott Fuller, Aug. 10, 1928, Fuller MSS, M1090:2:46:10).

24. Introduction to *4D Time Lock*, Fuller MSS, M1090:2:20:35, May 21, 1928.

25. Fuller to Vincent Astor, May 21, 1928, Fuller MSS, M1090:2:20:35.

26. "Cities Becoming 'Peas of One Pod,' Architects Warn," *St. Louis Star*, May 17, 1928. One can hardly miss the overtones of the famous duel between Henry Van de Velde and Hermann Muthesius on "*Typisierung*" that took place at the Cologne exhibition of the German Werkbund in 1914. For the texts see Ulrich Conrads, *Programs and Manifestoes on Twentieth Century Architecture* (Cambridge, MA: MIT Press, 1970), 28–31. For a nuanced reading of their epochal exchange see Frederic J. Schwartz, *The Werkbund: Design Theory and Mass Culture Before the First World War* (New Haven, CT: Yale University Press, 1996), 147–50. See also the classic formulation by Kenneth Frampton, "Towards a Critical Regionalism: Six Points for an Architecture of Resistance," in *The Anti-Aesthetic: Essays on Postmodern Culture*, ed. Hal Foster (Port Townsend, WA: Bay Press, 1983), 16–30.

27. Donald Deskey to Fuller, Feb. 25, Aug. 7, 1930, Fuller MSS, M1090:2:22:37; see also his acerbic remarks on "the myth of Industrial Design" in "The Comprehensive Man," in Fuller, *Ideas and Integrities*, 76–78. In a similar vein the Bauhaus would be curtly dismissed as the "foolish nonsense" of European stylists whose "superficial irrelevancies" did not touch underlying science (Fuller, "Influences on My Work," 27). On the Technocracy movement see Karl Markley Conrad, "Buckminster Fuller and the Technocratic Persuasion" (PhD diss., University of Texas at Austin, 1973); see also the scholarly appraisals of Carroll Pursell, *The Machine in America: A Social History of Technology* (Baltimore: Johns Hopkins University Press, 1995), 252–53; Edward T. Layton Jr., *The Revolt of the Engineers: Social Responsibility and the American Engineering Profession* (Baltimore: Johns Hopkins University Press, 1986), chap. 10.

28. [Harvard University office of the president] to R. Buckminster Fuller, May 28, 1928, Fuller MSS, M1090:2:20:34.

29. Bruce Barton to Lawrence Stoddard, June 8, 1928, Fuller MSS, M1090:2:20:34.

30. Fuller to J. M. Hewlett, May 21, 1928, Fuller MSS, M1090:2:46:7; Fuller to Buffington, Aug. 21, 1928, Fuller MSS, M1090:2:20:35.

31. Fuller to Victor Astor, Aug. 28, 1928, Fuller MSS, M1090:2:20:35.

32. In an intimate letter to Evelyn Schwartz (n.d. but presumably late 1931 or early 1932) he refers to "his nervous breakdown of the previous year" (Chronofile vol. 39 [1931]). Evelyn Stefansson Nef (née Schwartz) writes bitterly of her adolescent relationship with Fuller during that period in her autobiography, *Finding My Way* (Washington, DC: Francis Press, 2002), 47–58.

33. In an interview conducted in 1970 this charge was still able to evoke an intense reaction: "Don't talk to me about failure," he thundered back to Martin Pawley. "Failure is a word invented by men, there is no such thing as a failure in nature" (Pawley, *Buckminster Fuller*, 26).

34. The allusion here is to Jürgen Habermas's pivotal thesis on "the unfinished project of modernity." See Habermas, *The Philosophical Discourse of Modernity: Twelve Lectures*, trans. Frederick G. Lawrence (Cambridge, MA: MIT Press, 1991).

35. Hugh Kenner, *Bucky: A Guided Tour of Buckminster Fuller* (New York: Morrow, 1973), 156.

36. Fuller, "Later Development of My Thought," 42 (Fuller's emphasis).

37. Kenner, *Bucky*, 200.

Chapter 3

1. See Merritt Roe Smith and Leo Marx, eds., *Does Technology Drive History?* (Cambridge, MA: MIT Press, 1994).

2. See Howard P. Segal, "Eighteenth-Century American Utopianism: From the Potential to the Probable," *Utopian Studies* 11 (2000): 5–13.

3. R. Buckminster Fuller, *Grunch of Giants* (New York: St. Martin's, 1983), xiv.

4. For an early analysis of just the Tofflers and Naisbitt and Aburdene, see "High Tech and the Burden of History," chap. 12 in Howard P. Segal, *Future Imperfect: The Mixed Blessings of Technology in America* (Amherst: University of Massachusetts Press, 1994), 163–202.

5. See Jean Gimpel, *The Medieval Machine: The Industrial Revolution of the Middle Ages* (New York: Holt, Rinehart and Winston, 1976).

6. Fuller, *Utopia or Oblivion: The Prospects for Humanity* (New York: Overlook Press, 1969), 289–90, 292.

7. See my review of R. Buckminster Fuller, *Inventions: The Patented Works of R. Buckminster Fuller* (New York: St. Martin's, 1983); and of James Ward, ed., *The Artifacts of Buckminster Fuller: A Comprehensive Collection of His Designs and Drawings* (New York: Garland, 1985), in *Technology and Culture* 28 (July 1987), 697–98. Both books were posthumous publications.

8. To his credit, Gerard O'Neill makes no claims that his societies witness the literal perfection of the species and admits that human beings are by nature flawed. His colonies offer opportunities to pursue happiness but do not guarantee it. See his *The High Frontier: Human Colonies in Space* (Garden City, NY: Doubleday, 1982), 272–73; and his *2081: A Hopeful View of the Human Future* (New York: Simon and Schuster, 1981), 61–75.

9. See "Lewis Mumford's Alternatives to the Megamachine: Critical Utopianism, Regionalism, and Decentralization," in Segal, *Future Imperfect*, 152.

10. This paragraph and its quotations come from The Center for Land Use Inter-

pretation, available at http://ludb.clui.org/ex/i/CO3134/ (accessed Oct. 1, 2008); see also Timothy Miller, "The Farm: Roots of Communal Revival, 1962–1966," available at www.thefarm.org/lifestyle/root2.html (accessed Oct. 1, 2008).

11. Mumford, *Findings and Keepings: Analects for an Autobiography* (New York: Harcourt Brace Jovanovich, 1975), 373.

12. See the still very illuminating comparison by architectural critic Allan Temko, "Which Guide to the Promised Land? Fuller or Mumford?" in *Horizon* 10 (summer 1968): 25–30.

13. See Simon Ramo, *Cure for Chaos: Fresh Solutions to Social Problems Through the Systems Approach* (New York: David McKay, 1969); and David Halberstam, *The Best and the Brightest* (New York: Random House, 1972).

14. "Star Wars" is commonly understood in its basic structure and operation. On the far less familiar supercollider see "Preface, 1995: The Death of the Superconducting Super Collider in the Life of American Physics," in Daniel J. Kevles, *The Physicists: The History of a Scientific Community in Modern America* (1977; repr. Cambridge, MA: Harvard University Press, 1995), ix–xlii.

15. For an extended discussion of the fate of the Technocracy movement see my new introduction to Harold Loeb's *Life in a Technocracy: What It Might Be Like* (Syracuse, NY: Syracuse University Press, 1996; orig. published 1933).

16. Fuller, *Inventions*, xxxii, xx.

Chapter 4

1. See R. Buckminster Fuller, *Your Private Sky: R. Buckminster Fuller, Discourse*, ed. Joachim Krausse and Claude Lichtenstein (Baden: Lars Müller, 2001), 84–103; hereafter *Your Private Sky: R. Buckminster Fuller, Discourse* is referenced as vol. 2 of *YPS*.

2. R. Buckminster Fuller, *Education Automation: Freeing the Scholar to Return to His Studies* (Carbondale: Southern Illinois University Press, 1962), 8.

3. R. Buckminster Fuller, *Synergetics: Explorations in the Geometry of Thinking*, 2 vols. (New York: Scribner, 1975–79).

4. R. Buckminster Fuller, "Omnidirectional Halo," in *No More Secondhand God and Other Writings* (Carbondale: Southern Illinois University Press, 1963), 137; see also *Synergetics*, 1:237.

5. *Synergetics*, 2:31.

6. Ibid., 2:127.

7. Ibid., 2:31.

8. E. J. Applewhite, *Cosmic Fishing: An Account of Writing Synergetics with Buckminster Fuller* (New York: Macmillan, 1977), 47.

9. The author has edited three books with writings of R. Buckminster Fuller in German translation between 1968 and 1973.

10. William Kuhns, *The Post-Industrial Prophets* (New York: Weybright and Talley, 1971), 220.

11. R. Buckminster Fuller, *Cosmography*, ed. Kioshi Kuromiya (New York: Macmillan, 1992).

12. See Gerald Holton, *Einstein, die Geschichte und andere Leidenschaften* [Einstein, History and Other Passions] (Braunschweig, Wiesbaden: Vieweg-Verlag, 1998), 4:107.

13. See Harold W. Kroto, "Macro, Micro, and Nano-scale Engineering," in *Buckminster Fuller: Anthology for the New Millennium*, ed. Thomas K. Zung (New York: St. Martin's, 2001), 226.

14. Harold W. Kroto, "Die Entdeckung der Fullerene," in *Von Fuller bis zu Fullerenen*, ed. Wolfgang Krätschmer and Heike Schuster (Braunschweig, Wiesbaden: Verlag Vieweg, 1996), 59–60; Joachim Krausse, "Buckminster Fuller und seine Modellierung der Natur," in ibid., 25–51.

15. R. Buckminster Fuller, *Ideas and Integrities: A Spontaneous Autobiographical Disclosure*, ed. Robert Marks (Englewood Cliffs, NJ: Prentice-Hall, 1963), 75–76.

16. See Hugh Kenner, *Bucky: A Guided Tour of Buckminster Fuller* (New York: Morrow, 1973), 7–8.

17. See Reyner Banham, *Theory and Design in the First Machine Age* (London: Architectural Press, 1960), chap. 22.

18. Reyner Banham, "The Dymaxicrat" (1936), in *A Critic Writes: Essays by Reyner Banham* (Berkeley: University of California Press, 1996), 93–94.

19. See Kenner, *Bucky*, 129.

20. See R. Buckminster Fuller, *Your Private Sky: R. Buckminster Fuller, the Art of Design Science*, ed. Joachim Krausse and Claude Lichtenstein (Baden: Lars Müller, 1999), 80–87; hereafter *Your Private Sky: R. Buckminster Fuller, the Art of Design Science* is referenced as vol. 1 of *YPS*; see also *YPS* 2:62–74.

21. See *YPS* 1:114–15, 126–27, 134; *YPS* 2:92–96.

22. Compare the illustrations in *YPS* 1:83.

23. *YPS* 1:78–79.

24. R. Buckminster Fuller, "Vertical Is to Live, Horizontal Is to Die," in *YPS* 2.

25. R. Buckminster Fuller, "Lightful Houses" (1928), in *YPS* 2:70.

26. See Joachim Krausse and Claude Lichtenstein, "Earthwalking-Skyriding: An Invitation to Join Buckminster Fuller on a Voyage of Discovery," in *YPS* 2:23–31; R. Buckminster Fuller, "Noah's Ark #2," in *YPS* 2:176.

27. Anne Hewlett Fuller, "Diary starting August 7th when I came out to Chicago with Bucky," Dymaxion Chronofile, vol. 30 (1927–28), R. Buckminster Fuller Collection, Stanford University Libraries (hereafter cited as Fuller MSS).

28. See illustration in *YPS* 1:84.

29. R. Buckminster Fuller, "Lightful Houses," in *YPS* 2:70 (Fuller's emphasis).

30. R. Buckminster Fuller, "Mistake Mystique" (1979), in *YPS* 2:296.

31. See images in *YPS* 1:85–87; *YPS* 2:63.

32. Fuller, "Lightful Houses."

33. R. Buckminster Fuller, *Nine Chains to the Moon* (Philadelphia: J. P. Lippincott, 1938), 43.

34. See Reyner Banham, *The Architecture of the Well-tempered Environment* (London: Architectural Press, 1969); Joachim Krausse, "Environment Controlling—für eine Welt der Vielen: Buckminster Fullers Wirkung in Großbritannien," in *Norman Foster: Architecture Is About People*, ed. Susanne Anna (Köln: Museum of Applied Arts, 2002), 97.

35. R. Buckminster Fuller, *Designing a New Industry* (Wichita: Fuller Research Institute, 1946), 35.

36. R. Buckminster Fuller, "The 90% Automatic Cotton Mill" (1951), in *YPS* 1:336–39.

37. Fuller, *Designing a New Industry*, 35.

38. Ibid.

39. Ludwig Bertalanffy, *General System Theory* (New York: George Braziller, 1968).

40. See Joachim Krausse, "Architektur der Hochtechnologie: Buckminster Fullers Dymaxion Haus 1929," in *1929: Beiträge zur Archäologie der Medien*, ed. Stefan Andriopoulos and Bernhard J. Dotzler (Frankfurt/Main: Suhrkamp Verlag, 2002), 192–223.

41. R. Buckminster Fuller to Rosamond Fuller, June 13, 1928, in *4D Time Lock* (1928) (repr. Albuquerque: Lama Foundation, 1972), 79, quoted in *YPS* 1:77.

42. R. Buckminster Fuller, *Ideas and Integrities: A Spontaneous Autobiographical Disclosure* (Englewood Cliffs, NJ: Prentice-Hall, 1963), 69.

43. Le Corbusier, *Towards a New Architecture*, trans. Frederick Etchells (London: John Rodker, 1927).

44. R. Buckminster Fuller, 4D, copy no. 1, Appendix 1, References, manuscript 5 pp. Dymaxion Chronofile, vol. 35 (1928), Fuller MSS.

45. Fuller, *Ideas and Integrities*, 70.

Chapter 5

1. Fuller's first name for his housing project was "Lightful Houses" (Feb. 1928); he altered it to "4D" (spring 1928). In 1929 a Chicago advertisement specialist proposed to Fuller the keyword *Dymaxion*, which Fuller used until ca. 1950. These projects were all closely related, encompassing a general progression and evolution of a housing project that developed over several decades.

2. Fuller mentioned later that he had ideally foreseen a circular plan for the 4D house, but he replaced that plan owing to reasons of viability (bending of metal tubes) by a hexagonal scheme; see R. Buckminster Fuller, *Designing a New Industry* (Wichita: Fuller Research Institute, 1946), 22.

3. R. Buckminster Fuller, *4D Time Lock* (1928; repr. Albuquerque, NM: Lama Foundation, 1970).

4. Fuller, *Designing a New Industry*, 16.

5. R. Buckminster Fuller, *Your Private Sky: R. Buckminster Fuller, Discourse*, ed. Joachim Krausse and Claude Lichtenstein (Baden: Lars Müller, 2001), 48; hereafter referenced as vol. 2 of *YPS*.

6. Ibid.

7. What is coined "international style" in the Anglo-Saxon countries is in German-speaking countries "neues Bauen" (or "neue Sachlichkeit"); in France "la construction nouvelle"; in Italy "l'architettura razionale."

8. *ABC—Beiträge zum Bauen* (Basel: Thalwill, 1926), 1:8.

9. R. Buckminster Fuller, "Doing the Most with the Least," *Shelter*, no. 4 (1932): 36; depicted in R. Buckminster Fuller, *Your Private Sky: R. Buckminster Fuller, the Art of Design Science*, ed. Joachim Krausse and Claude Lichtenstein (Baden: Lars Müller, 1999), 166; hereafter referenced as vol. 1 of *YPS*.

10. "Das Bauen hat aufgehört, eine Sache der Kunst zu sein, es hat überall . . . die Arbeitsweise der Technik übernommen, die eindeutig aus dem Material und seinen

Möglichkeiten, aus dem Vorgang des Bauens aus den Anforderungen an das fertige Bauwerk heraus gestaltet." [Building is no more an artistic issue; it has thoroughly . . . adopted the methods of engineering which require to design according to the materials and their potentials and along the building process, obeying the demands of the planned building.] Hans Schmidt, "Das Bauen ist nicht Architektur," *Werk* (Zurich), May 1927, 141 (my translation).

11. R. Buckminster Fuller, "Dymaxion House," lecture presented to the Architectural League, New York City, July 1929, in *YPS* 2:84–103.

12. *Fortune*, July 1932; repr. in *YPS* 1:136–37.

13. This aspect is discussed by Federico Neder in his doctoral dissertation on R. Buckminster Fuller, submitted in Geneva in 2003 (Ecole d'Architecture, unpublished).

14. The name "Stockade" for Hewlett's and Fuller's building system seems to fit this interpretation.

15. *YPS* 2:87.

16. André Corboz, "'Non-City' Revisited," in *Die Kunst, Stadt und Land zum Sprechen zu bringen* (Basel: Birkhäuser, 2001), 123–39.

17. R. Buckminster Fuller, *Nine Chains to the Moon* (Philadelphia: J. P. Lippincott, 1938), 34.

18. Joachim Krausse suggests that the label "4D" might possibly also refer to (Henry) Ford.

19. For Fuller's complete array of arguments and exhaustive description of his project see the transcription of his 1929 lecture delivered at the Architectural League, New York City (*YPS* 2:84–103).

20. R. Buckminster Fuller, *No More Secondhand God* (Carbondale: Southern Illinois University Press, 1970), 52.

21. Fuller, *4D Time Lock*, 18.

22. Le Corbusier, "Appel aux industriels" (1924), in Le Corbusier and Pierre Jeanneret, *Oeuvre complète*, vol. 1 (1910–29) (Zurich: Girsberger, 1929), 77.

23. Catharine Beecher and Harriet Beecher Stowe, *The American Woman's Home* (1869; repr. Hartford, CT: Stowe-Day Foundation, 1975), 25.

24. Ibid., 58.

25. See *YPS* 1:239–45.

26. Beecher and Beecher Stowe, *The American Woman's Home*, 43.

27. Ibid., 49. See also Reyner Banham, *Architecture and the Well-tempered Environment* (London: Architectural Press, 1969), 96–99. Banham correlated Beecher's design with Fuller's, emphasizing the physical aspect of climate control. The Dymaxion house was just mentioned, not discussed.

28. Fuller, *4D Time Lock*, 12.

29. R. Buckminster Fuller, "Can't Fool Cosmic Computer," from *Grunch of Giants*, repr. in Thomas T. K. Zung, *Buckminster Fuller: Anthology for the New Millennium* (New York: St. Martin's, 2001), 275.

Chapter 6

1. For an overview see David E. Nye, *Consuming Power: A Social History of American Energies* (Cambridge, MA: MIT Press, 1998), 217–48.

2. R. Buckminster Fuller, "Guinea Pig B," in Thomas T. K. Zung, *Buckminster Fuller: Anthology for the New Millennium* (New York: St. Martin's, 2001), 312.

3. R. Buckminster Fuller, interview transcript from "Energy and Technology," WOSU-AM-FM-TV, April 2, 1979, R. Buckminster Fuller Collection, M1090:8:182:2:29, 31–32, Stanford University Libraries (hereafter cited as Fuller MSS).

4. The classic discussion on the problems of the Newtonian synthesis is in Thomas Kuhn, *The Structure of Scientific Revolutions*, 2nd ed. (Chicago: University of Chicago Press, 1970). See also Sherry H. Olson, *The Depletion Myth: A History of Railway Use of Timber* (Cambridge, MA: Harvard University Press, 1971).

5. Nikola Tesla, "The Problem of Increasing Human Energy with Special Reference to the Harnessing of the Sun's Energy," *Century Magazine*, May 1900, 175.

6. R. Buckminster Fuller, *Synergetics* (New York: Macmillan, 1975), 258.

7. Ibid., 256.

8. R. Buckminster Fuller, *Utopia or Oblivion: The Prospects for Humanity* (Harmondsworth: Penguin, 1969).

9. R. Buckminster Fuller, *Critical Path* (New York: St. Martin's, 1981), xviii.

10. R. Buckminster Fuller, keynote speech, Conference on Energy and the Future of America's Communities, Fuller MSS, M1090:8:181:2:1.

11. Ibid.

12. Ibid., 9. See also Fuller, *Utopia or Oblivion*, 178–79; and R. Buckminster Fuller, "The Age of Astro Architecture," *Saturday Review*, July 13, 1968.

13. Fuller, keynote speech, 10.

14. R. Buckminster Fuller, "Comprehensive Designing," in Zung, *Buckminster Fuller*, 73.

15. Glenn A. Olds, "R. Buckminster Fuller: Cosmic Surfer," in ibid., 61.

16. Transcripts of R. Buckminster Fuller's comments at the International Conferences on Environmental Future, Sep. 1977, Fuller MSS, M1090:8:176:4.

17. Fuller, *Synergetics*, 533.

18. See James Miller, ed., *The Buckminster Fuller Reader* (London: Jonathan Cape, 1970), 38–41.

19. Horatio Greenough, *Form and Function: Remarks on Art, Design, and Architecture* (Berkeley: University of California Press, 1958), 116–29. On Henry Ford and Ralph Waldo Emerson see David E. Nye, *Henry Ford: "Ignorant Idealist"* (New York: Academic Press, 1979), 93–95, 106.

20. "Frank Lloyd Wright and *Nine Chains to the Moon*," in Zung, *Buckminster Fuller*, 366–68.

21. Fuller, *Critical Path*, 253. "No single move can bring us more swiftly in the direction of complete overall desovereignization, unblocking the free flow of technologies and resources, than that of instituting the world-around integrated electrical-energy network" (ibid., 309).

22. Charles Steinmetz, "Effect of Electrical Engineering on Modern Industry," *Scientific American*, suppl. no. 2002, vol. 77, May 16, 1914, 314.

23. www.geni.org/globalenergy/issues/overview/grid.shtml (accessed Dec. 20, 2008).

24. On Fuller's involvement with Ohio University students see transcription of conversation between Fuller and John McHale, Fuller MSS, M1090:8:19:5:6.

25. On this industrial tourism see David E. Nye, *American Technological Sublime* (Cambridge, MA: MIT Press, 1994), chap. 5.

26. Jeffrey Meikle, *Design in America* (New York: Oxford University Press, 2005), 117.

27. The restored Dymaxion house can be seen at www.hfmgv.org/museum/dymaxion.aspx (accessed Dec. 20, 2008).

28. An investor, William Gordon, purchased the only two prototypes of this round, aluminum house, and in 1948 he inserted one of them into his family's home, where it remained until ultimately bought, restored, and exhibited in the Henry Ford Museum.

29. Fuller, *Utopia or Oblivion*, 410.

30. Ibid., 412–13.

31. Fuller, *Critical Path*, 315.

32. Buckminster Fuller received a U.S. patent for the geodesic dome. However, at least one dome of a similar nature preceded his. Dr. Walter Bauersfeld of the Carl Zeiss Optical Company in Germany built between 1919 and 1922 a hollow globular structure made from 3,480 metal struts that were pieced together as a series of triangles. It was then covered with a thin layer of concrete for use as a planetarium with Zeiss projection equipment. The design proved successful and was widely adopted. One, the famous Adler Planetarium, was erected in Chicago in 1930. See www.telacommunications.com/geodome.htm (accessed Dec. 20, 2008); and Werner Helmet, *From the Arratus Globe to the Zeiss Planetarium* (Stuttgart: Gustav Fischer, 1957). In addition, elements of Fuller's work were independently anticipated by Alexander Graham Bell when he was building kites and airplane wings. See Dorothy Harley Eber, prologue to *Genius at Work: Images of Alexander Graham Bell* (New York: Viking, 1982).

33. Zung, *Buckminster Fuller*, 39.

34. Fuller, *Critical Path*, 317–22.

35. Ibid., 320.

36. Sylvia Hart Wright, *Sourcebook of Contemporary North American Architecture: From Postwar to Postmodern* (New York: Van Nostrand Reinhold, 1989), 33.

37. "Geodesic Structures," from *The Dymaxion World of Buckminster Fuller*, repr. in Zung, *Buckminster Fuller*, 40.

Chapter 7

1. The seminal instance of this approach is Reyner Banham, *Theory and Design in the First Machine Age* (London: Architectural Press, 1960).

2. The *4D Time Lock* manuscript can be found in R. Buckminster Fuller Collection, M1090:8:1–2, Stanford University Libraries (hereafter cited as Fuller MSS). Privately published in a mimeographed edition of approximately two hundred copies in 1928, the treatise was reissued, with the addition of drawings Fuller made during 1927 and 1928, as R. Buckminster Fuller, *4D Time Lock* (Albuquerque, NM: Lamas Foundation, 1972). Many of the ideas in *4D Time Lock* first appeared in a manuscript titled "Lightful Houses" (1928), Fuller MSS, M1090:8:2.

3. R. Buckminster Fuller, *Nine Chains to the Moon*, rev. ed. (Garden City, NY: Anchor, 1971), 89.

4. Calvin Tomkins, "In the Outlaw Area," *New Yorker*, Jan. 8, 1966, 35–97, 40.

5. For a survey of Western sumptuary regulation see Alan Hunt, *Governance of the Consuming Passions: A History of Sumptuary Law* (New York: St. Martin's, 1996).

6. See Jonathan Massey, "New Necessities: Modernist Aesthetic Discipline," *Perspecta: The Yale Architectural Journal* 35 (2004): 112–33.

7. R. Buckminster Fuller, with Kiyoshi Kuromiya, *Critical Path* (New York: St. Martin's, 1981), chap. 4, "Self-Disciplines of Buckminster Fuller."

8. R. Buckminster Fuller, *Ideas and Integrities: A Spontaneous Autobiographical Disclosure*, ed. Robert W. Marks (New York: Collier, 1963), 30; see also Lloyd Steven Sieden, *Buckminster Fuller's Universe* (Cambridge, MA: Perseus, 2000), chap. 4, "Years of Deterioration."

9. See Fuller to Anne Hewlett Fuller, July 31, 1932, Fuller MSS, M1090:2:47; and Sieden, *Buckminster Fuller's Universe*, 87.

10. Fuller, "Lightful Houses," 59, 65.

11. Fuller, *4D Time Lock*, 1.

12. Fuller, "Motion Economics and Contact Economies," unpublished TS (1942), Fuller MSS, M1090:8:5. "Technocracy aspires to exercising political power," Fuller explained to Technocracy, Inc., member Tom Gibbins in 1976, "in contradistinction to what I call exclusive preoccupation with reducing to practice, and by proving and developing artifacts by comprehensive design science initiatives" (Fuller to Gibbins, Dec. 12, 1976, Fuller MSS, M1090:2:330).

13. Fuller, *Nine Chains to the Moon*, 175.

14. In the 1970s, for instance, Fuller described the energetic-synergetic geometry he developed with E. J. Applewhite as the only coordinate system capable of representing nature's four spatial dimensions. Dismissing the temporal conception of the fourth dimension as a myth developed by scientists to avoid confronting their incapacity to model higher-dimensional shapes, Fuller gave energetic-synergetic geometry four axes at sixty degree angles to one another, like the four sides of a tetrahedron. According to Fuller, this allowed the system to model "fourth-order problems," four-dimensional shapes that could not be modeled using a three-dimensional Cartesian geometry. See Fuller, *Utopia or Oblivion: The Prospects for Humanity* (New York: Bantam, 1969), 84, 97. The tetrahedron, he explained in another context, was "the fundamental minimal structure of the universe" because "nature is inherently four dimensional and sometimes more" (Fuller to Linda Dalrymple Henderson, Oct. 2, 1979, quoted in Linda Dalrymple Henderson, *The Fourth Dimension and Non-Euclidean Geometry in Modern Art* [Princeton, NJ: Princeton University Press, 1983], 235). In his early work, by contrast, Fuller more often emphasized the temporal fourth dimension. In 1928, for instance, he characterized "4D" as "only the enigmatic term for time" ("Time Lock" MS, 39, in Fuller MSS, M1090:8:1).

15. Riemann's 1854 lecture "Über die Hypothesen, welche der Geometrie zu Grunde liegen," which first presented his theory of *n*-dimensional space, was published in German in 1868, then in English in 1873. See Max Jammer, *Concepts of Space: The History of Theories of Space in Physics*, 2nd ed. (New York: Dover, 1969), esp. 151–52.

16. Since the 1970s, however, Riemann space has enjoyed a resurgence owing to its use in superstring theory, which speculates that as many as ten dimensions of space, along with an eleventh consisting of time, might make up our universe. See Brian

Greene, *The Elegant Universe* (New York: Vintage, 2000); and Brian Greene, *The Fabric of the Cosmos* (New York: Vintage, 2005).

17. See Fuller, *Nine Chains to the Moon*, 185. Fuller also used the fourth dimension to link time and space. He believed that all bodies were essentially spherical, their deformations from that ideal shape being produced by the pressure of their interaction with other adjacent bodies. "Everything is round in its perfect state," he explained in an early *Time Lock* draft, even if it could be compressed and distorted by contact with other bodies much like a pneumatic tire encountering a rock or other object. These spherical bodies expanded radially over time, and the rate of this "radiation" helped explain the relative size of things. "*Matter takes time*," he stated. "That's 4th Dim" ("Lightful Houses," 65). Fuller used the term *relativity* to describe the varying "wave length and frequency" of this expansion along the "radiant time dimension," which manifested itself to normal human perception as growth and longevity ("Lightful Houses," 44). See also Fuller, *Nine Chains to the Moon*, 117–18; and Fuller, *Time Lock*, 36–38.

18. In June 1928 Fuller discussed his understanding of the fourth dimension with his father-in-law, James Monroe Hewlett. "After I had worked out my own *time laws* of relativity," he explained, "I decided to study the books by Albert Einstein." Simultaneously claiming that Einstein's ideas corresponded closely to his own conceptions and admitting that he found them "obscurantist," Fuller concluded that "Time and Relativity are essential components of CONSTRUCTION DESIGN and HARMONIC COMPOSITION. . . . The abstract starting point must be consistently adhered to in the complete subjection of materialism to the will of the unselfish or spiritual man" (Fuller to James Monroe Hewlett, June 28, 1928, transcribed in *Time Lock*, 5–6).

19. Fuller, *Nine Chains to the Moon*, 134.

20. Fuller, in *4D; Buckminster Fuller's Dymaxion House*, exhibition catalog (Cambridge, MA: Harvard Society for Contemporary Art, June 1929), n.p.

21. Fuller, *Nine Chains to the Moon*, 181, 185. In a further application of relativity theory to scientific management, Fuller compared the various kinds of "therblig," the sixteen maximally efficient human work actions identified in 1919 by Taylorist efficiency engineers Frank and Lillian Gilbreth, to Einstein's c^2, "i.e., TOP *efficient* speed" (Fuller, *Nine Chains to the Moon*, 80).

22. Fuller, *Time Lock*, 36. "In consideration of the fact that no matter can exist without *time*, else it would not exist; and that the *time* dimension is the most important dimension of all matter; and that all our industry is but a *time* saving institution; that all sport is but a *time* controlling demonstration; and that all art is but an harmonic division, composition, and projection of *time*; and that we are fast approaching a *time* standard (men dollar hours) instead of a gold standard. . . . In full consideration of this new economic law must the new era home be designed and its plans of industrialization evolved," Fuller asserted.

23. For Taylor's outline of scientific management practice see Frederick W. Taylor, *The Principles of Scientific Management* (New York: Harper and Brothers, 1911). Regarding Taylor's "one best way" philosophy see Robert Kanigel, *One Best Way: Frederick Winslow Taylor and the Enigma of Efficiency* (New York: Viking, 1997).

24. Fuller, *Time Lock*, 51.

25. For an overview of the Technocracy movement see William E. Akin, *Technocracy*

and the American Dream: The Technocrat Movement, 1900–1941 (Berkeley: University of California Press, 1977). After continuing to develop through the 1920s, technocracy gained widespread attention in 1932, when the Depression made the concept of rational planning by disinterested experts seem like an urgently needed alternative to laissez-faire capitalism. Although 1933 and 1934 saw an explosion of societies and chapters dedicated to technocratic ideals, the movement rapidly lost momentum to Roosevelt's New Deal reform program, which co-opted many of its arguments.

26. Akin, *Technocracy and the American Dream*, 10.

27. Technocracy, Inc., *Technocracy Study Course*, 5th ed. (New York: Technocracy, 1940 [1934]), vii. See also Akin, *Technocracy and the American Dream*, 34–41.

28. Veblen's 1919 *Dial* essays were collected in his book *The Engineers and the Price System* (New York: B. W. Huebsch, 1921).

29. Fuller to John Omdahl, Sep. 16, 1970, in Fuller MSS, M1090:2:205. See also Fuller to Gibbins, Dec. 12, 1976.

30. Fuller's work on *Shelter* was supported by a thousand-dollar grant from the Architects' Emergency Relief Committee of New York, the same organization that funded Scott's energy survey. See Fuller to Gibbins, Dec. 12, 1976; and Akin, *Technocracy and the American Dream*, 62.

31. Fuller, *Nine Chains to the Moon*, 88–89. Fuller later recalled that he had diverged from the organization when it "rather alarmingly adopted blue military uniforms and blue-grey Technocracy staff cars which were reminiscent of other colored shirt movements of the Nazis" (Fuller to Gibbins, Dec. 12, 1976); see also Fuller to Omdahl, Sep. 16, 1970.

32. *Shelter*, vol. 2, bulletin 3, June 22, 1932, 88.

33. Fuller, "Motion Economics and Contact Economies," unpublished TS (1942), in Fuller MSS, M1090:8:5. "Technocracy aspires to exercising political power," Fuller explained to Technocracy, Inc., member Tom Gibbins in 1976, "in contradistinction to what I call exclusive preoccupation with reducing to practice, and by proving and developing artifacts by comprehensive design science initiatives" (Fuller to Gibbins, Dec. 12, 1976).

34. Fuller, *Nine Chains to the Moon*, 195; and Tomkins, "In the Outlaw Area," 64. In discussing the aims and methods of his work, Fuller frequently cited as a model the nautical "trim tab," a small flap on a ship's rudder that initiates steering motions. Like the tiny trim tab, the "unsupported individual" who lacked—or rejected—the authority of a command structure could best stimulate large-scale change by establishing incentives for people to modify their behavior. "In a dynamic universe," Fuller would later explain, "everything is always moving in the direction of least resistance. . . . If that's the case, then it should be possible to modify the shapes of things so that they follow preferred directions of least resistance." For a longer discussion of the "new forms rather than reforms" principle see R. Buckminster Fuller and John McHale, *World Design Science Decade 1965–1975 Phase I (1963) Document 1: Inventory of World Resources Human Trends and Needs* (Carbondale: Southern Illinois University Press, 1963), Appendix A, 51–58.

35. Quoted in Tomkins, "In the Outlaw Area," 85.

36. Walter Peters, "Buckminster Fuller and the Indlu Geodesic Dome Project, South Africa, 1958," paper presented at the 58th annual meeting of the Society of Architectural Historians, Vancouver, April 9, 2005.

37. See the report on the 90 percent Automatic Cotton Mill studio Fuller conducted at North Carolina State College, Jan. 2 to Feb. 2, 1952, grouped with other student project files in Fuller MSS, M1090:18:107.

38. In this he followed nineteenth-century house reformer Orson Fowler, who had promoted octagon houses in similar terms. On Fowler and Fuller see Michael J. Auer, "The Dymaxion Dwelling Machine," in H. Ward Jandl, *Yesterday's Houses of Tomorrow* (Washington, DC: Preservation Press, 1991), 83–99.

39. For a comparison of the 4D/Dymaxion house to other prefabricated housing proposals see Alfred Bruce and Harold Sandbank, *A History of Prefabrication* (New York: John B. Pierce Foundation, July 1943).

40. Fuller, "Lightful Houses," 50.

41. Fuller, *Nine Chains to the Moon*, 17.

42. Fuller to George N. Buffington, Aug. 31, 1928, incorporated into *Time Lock*, 120–48.

43. Fuller, *Time Lock*, 123.

44. Fuller, "Lightful Houses," 56–58. "As time is saved by individuals for other individuals, so shall that individual control capital in the direct equivalent of man hours saved, less cost of equipment and its maintenance," Fuller explained. "But inversely . . . having proven his (or her) worth per hour in time saving, when he indulges the bestial stomach so shall he lose capital control proportional per hour and as he influences and wastes other individuals time thereby." Using Henry Ford as an example, Fuller outlined how his system might operate in practice. Assuming that Ford was worth a million dollars a day to the world, he should be compensated at that rate for all time spent sleeping, eating, exercising, and cleaning himself. Every minute Ford spent "in self indulgence" such as gossip, however, would be debited from his compensation.

45. See Claude Bragdon, *Projective Ornament* (Rochester NY: Manas Press, 1915). For a comprehensive account of Bragdon's work see Jonathan Massey, *Crystal and Arabesque: Claude Bragdon, Ornament, and Modern Architecture* (Pittsburgh: University of Pittsburgh Press, 2009).

46. Bragdon also employed another technique for generating geometric patterns: tracing "magic lines," the lines created by tracing in ascending numerical order the numbers in a "magic square," an arrangement of sequential numbers into a square, each column, row, and diagonal of which sums to the same number.

47. Bragdon, *Projective Ornament*, 63–64.

48. See, e.g., Bragdon, "New Concepts of Time and Space," *The Dial*, Feb. 1920, 187–91.

49. Fuller, *Time Lock*, 46. Bragdon's copy is listed as number 66 in the handwritten list "Registration of Copies" included in the Chronofile, vol. 35 (1928), in Fuller MSS, M1090:2:20.

50. Bragdon, *Projective Ornament*, 64.

51. See Fuller, *Nine Chains to the Moon*, 18–22; and Bragdon, *Four-Dimensional Vistas* (1914), 2nd ed. (New York: Alfred A. Knopf, 1923), 116.

52. Fuller, *Time Lock*, 35.

53. Fuller, "Ballistics of Civilization," MSS (1939), in Fuller MSS, M1090:8:4.

Chapter 8

1. Cranston Jones to Buckminster Fuller, Sep. 23, 1959; in R. Buckminster Fuller Collection, M1090:2:106, Stanford University Libraries (hereafter cited as Fuller MSS).

2. Based on corrugated iron grain-storage bins, the *Dymaxion Deployment Unit* (*DDU*) was a prefabricated bomb shelter or "defense house" comprising two cylindrical metal units; see Beatriz Colomina, "DDU at MoMA," *ANY* magazine, no. 17 (Jan. 1997): 49–53.

3. "By erecting in its Garden from time to time exhibition houses or demonstration structures, the Museum has been able to make clear to the layman many points on architecture, design and construction" (*Today and Tomorrow* [New York: Museum of Modern Art, 1960], 22; in Fuller MSS, M1090:3:6).

4. Peter Blake, *No Place like Utopia: Modern Architecture and the Company We Kept* (New York: Alfred A. Knopf, 1993), 138.

5. "No. 72 for Release Friday, August 28, 1959," in [Press Releases] The Museum of Modern Art Library, New York, volume for 1959 (hereafter cited as MoMA Library).

6. Arthur Drexler, "Introductory Label: Structures by Buckminster Fuller," n.d. [1959], 2, in [Press Releases] MoMA Library, volume for 1959.

7. Arthur Drexler, *Three Structures by Buckminster Fuller in the Garden of the Museum of Modern Art, New York* (New York: Museum of Modern Art, 1960), n.p.

8. Arthur Drexler, "Text of Wall Label: Structures by Buckminster Fuller," n.d. [1959], in [Press Releases] MoMA Library, volume for 1959.

9. Ibid.; Anon., "R. Buckminster Fuller," n.d. [1959], in [Press Releases] MoMA Library, volume for 1959.

10. Yunn Chii Wong, "The Geodesic Works of Richard Buckminster Fuller, 1948–68 (The Universe as Home of Man)," 2 vols. (PhD diss., Massachusetts Institute of Technology, 1999), 1:334. Wong states that 550 geodesic rigid radomes were constructed between 1954 and 1961 and estimates Fuller's royalties at 5 percent of fabrication costs, giving a ballpark average of $2,000 in royalties per radome.

11. Drexler, "Text of Wall Label."

12. Anon., "Art Carnival on for Young," *New York Times*, Sep. 23, 1959, 34. "New Moods in Manhattan," *Town and Country*, March 1960, in Department of Public Information Records, 82 [21, 807], Museum of Modern Art Archives, New York (hereafter cited as MoMA Archives).

13. Anon., "R. Buckminster Fuller," in [Press Releases] MoMA Library, volume for 1959; and Drexler, "Text of Wall Label."

14. See installation photograph labeled on verso "Buckminster Fuller Exhibition: Far West Gallery. Oct. 27–Nov. 23, 1959. Photo: Kenneth Snelson, #10," in Department of Architecture and Design Exhibition Files, Exh. #652, MoMA Archives.

15. Drexler, *Three Structures.*

16. Ada Louise Huxtable, "Future Previewed? Innovations of Buckminster Fuller Could Transform Architecture," *New York Times*, Sep. 27, 1959, 21.

17. See Mark Wigley, "Planetary Homeboy," *ANY* magazine, no. 17 (Jan. 1997): 16.

18. *Visionary Architecture* opened on Sep. 29, 1960, thus running concurrently with the tail end of *Three Structures.*

19. Arthur Drexler, New York, to James W. Fitzgibbon, Raleigh, Feb. 15, 1960, in Fuller MSS, M1090:2:107.

20. Ibid. (emphasis added).

21. Buckminster Fuller to Elizabeth Shaw, New York, Oct. 21, 1959, in Fuller MSS, M1090:2:105.

22. Department of Architecture and Design Exhibition Files, Exh. #652, MoMA Archives.

23. See note 14.

24. Alex Soojung-Kim Pang, "Dome Days: Buckminster Fuller in the Cold War," in *Cultural Babbage: Technology, Time and Invention*, ed. Francis Spufford and Jenny Uglow (London: Faber and Faber, 1996), 169.

25. Arthur Drexler, New York, to J. A. Vitale, Lexington, Mass., June 8, 1959, in Fuller MSS, M1090:2:103. Vitale was division head at MIT's Lincoln Labs.

26. For details on Fuller's franchises and companies see Wong, "The Geodesic Works," 1:284–301. For a major such dispute see Jeffrey Lindsay, Montreal, to Buckminster Fuller, New York, March 7, 1955, in Fuller MSS, M1090:2:87.

27. See, e.g., Fuller MSS, M1090:2:107.

28. See Margaret Fitzgibbon, interview by Yunn Chii Wong, St. Louis, Missouri, Sep. 15, 1994, cited in Wong, "The Geodesic Works," 1:328n122.

29. Shoji Sadao, New York, to Shell Oil Company, Martinez, CA, Dec. 29, 1959, in Fuller MSS, M1090:2:105.

30. Bernie Kirschenbaum, interview by Yunn Chii Wong, New York, Oct. 5, 1994, cited in Wong, "The Geodesic Works," 1:328n122.

31. Ibid., 327–29.

32. See installation photograph labeled on verso "Buckminster Fuller Exhibition: Far West Gallery. Oct. 27–Nov. 23, 1959. Photo: Kenneth Snelson, #3," in Department of Architecture and Design Exhibition Files, Exh. #652, MoMA Archives.

33. Drexler, "Text of Wall Label." For Snelson's account of his own role in Drexler's eleventh-hour enlightenment about the origins of tensegrity see Kenneth Snelson, "Snelson on the Tensegrity Invention," *International Journal of Space Structures* 11, nos. 1–2 (1996): 46–47. For his account of the process of experimentation in which *Early X Piece* came about see Kenneth Snelson, New York, to René Motro, Guildford, Surrey, ca. 1990, in René Motro, *Tensegrity: Structural Systems for the Future* (London: Kogan Page, 2003), 225. Beyond the scope of the present chapter is the question of Karl Ioganson's invention of what might be called a prototensegrity principle in 1921; see Maria Gough, *The Artist as Producer: Russian Constructivism in Revolution* (Berkeley: University of California Press, 2005), 87–93. In a later forum I hope to reconsider this controversial question in the light of the argument presented here.

34. I have drawn these appellations from a letter by the late Wilder Green describing all the models loaned for the supplementary show; the four Snelson models were borrowed from the artist himself (see Wilder Green, New York, to Dorothy Dudley, Oct. 27, 1959, in Registrar Exhibition Files, Exhibition #649, MoMA Archives). While Snelson prefers not to use Fuller's term *tensegrity* in relation to his own work, this is nevertheless how his models were described in the preparation of the show. The photograph that appears at the rear of the vitrine was captioned "Tensegrity Dome

(1954), Fuller Student Project, University of Minnesota," a project in which Snelson was not involved.

35. "No. 97 for release Wednesday, October 28, 1959," in [Press Releases] MoMA Library, volume for 1959.

36. Anon., "Why Not Build It?" *New York Sunday News*, Nov. 1, 1959; Anon., "Exhibition Design for Future Living," *New York Post*, Nov. 11, 1959; George McCue, "Buckminster Fuller's One-Man Show," *St. Louis Post-Dispatch*, Nov. 1, 1959; in Department of Public Information Records, 82 (21, 807), MoMA Archives.

37. John Canaday, "Art: New Directions in Architecture," *New York Times*, Sep. 22, 1959, 78; Huxtable, "Future Previewed?" 21.

38. See "Geodesic Dome," *Architectural Forum* 95, no. 2 (Aug. 1951): 149.

39. Kenneth Snelson, New York, to René Motro, Guildford, Surrey, ca. 1990, in Motro, *Tensegrity*, 225.

40. Ibid., 223–24.

41. John Canaday, "Art: Constructions on the 'Tensegrity' Principle," *New York Times*, April 16, 1966, 24.

42. Kenneth Snelson, "Whose Baby?" Letter to the Editor, *Johns Hopkins Magazine*, April 1980, 2.

43. Kenneth Snelson, New York, to Buckminster Fuller, Jan. 23, 1972, in Fuller MSS, M1090:2:234.

44. Buckminster Fuller, Philadelphia, to Kenneth Snelson, New York, March 1, 1980, 24, in Fuller MSS, M1090:2:411. Fuller's account of the MoMA episode differs in certain details from that of Snelson.

45. Buckminster Fuller, "Tensegrity," *Portfolio and Art News Annual*, no. 4 (1961): 112–27, 144–48.

46. Wong, "The Geodesic Works," 1:168–69 and n. 72, where he cites Duncan Stuart, "Interview with Yunn Chii Wong," Raleigh, North Carolina, April 26, 1995. See also Michael John Gorman, *Buckminster Fuller: Designing for Mobility* (Milan: Skira, 2005).

47. Fuller, "Tensegrity," 121.

48. Buckminster Fuller to Brian Higgins, Chicago, Aug. 12, 1982, in Fuller MSS, M1090:2:465.

49. Pang, "Dome Days," 171.

50. Buckminster Fuller, Philadelphia, to Kenneth Snelson, New York, March 1, 1980, 4, 6, in Fuller MSS, M1090:2:411.

Chapter 9

1. Quoted in Kenneth Keniston, *Young Radicals: Notes on Committed Youth* (New York: Harcourt Brace and World, 1968), 48.

2. Spencer R. Weart, *Nuclear Fear: A History of Images* (Cambridge, MA: Harvard University Press, 1988), 133.

3. Elaine Tyler May, *Homeward Bound: American Families in the Cold War Era* (New York: Basic Books, 1988), 13–16.

4. C. Wright Mills, *The Power Elite* (New York: Oxford University Press, 1956), 3; quoted in Andrew Jamison and Ron Eyerman, *Seeds of the Sixties* (Berkeley: University of California Press, 1994), 42.

5. C. Wright Mills, *The Sociological Imagination* (New York: Oxford University Press, 1959), 168.

6. Ibid., 169.

7. Ibid., 171.

8. Quoted in May, *Homeward Bound*, 145.

9. See, e.g., Keniston, *Young Radicals*, 229–47.

10. Quoted in Keniston, *Young Radicals*, 39.

11. R. Buckminster Fuller, "The Comprehensive Designer," manuscript of 7 pages, June 1, 1949, Buckminster Fuller Archive, Manuscript File 49.06.01; repr. in R. Buckminster Fuller, *Your Private Sky: R. Buckminster Fuller, Discourse* (Baden: Lars Müller, 2001), 243–46; hereafter *Your Private Sky: R. Buckminster Fuller, Discourse* is referenced as vol. 2 of *YPS*. Fuller published an expanded version of this essay under the title "Comprehensive Designing," in *Trans/Formation* 1, no. 4 (1950): 18–23. This expanded version was reprinted in R. Buckminster Fuller, *Ideas and Integrities: A Spontaneous Autobiographical Disclosure*, ed. Robert Marks (Englewood Cliffs, NJ: Prentice-Hall, 1963), 173–82.

12. Fuller, *Ideas and Integrities*, 43.

13. Ibid., 35–43.

14. Ibid., 173.

15. Ibid., 176.

16. Ibid.

17. Ibid., 63.

18. Ibid., 249.

19. "When I heard that Aunt Margaret said, 'I must start with the universe and work down to the parts, I must have an understanding of it,' that became a great drive for me," he would later recall (R. Buckminster Fuller and Robert Snyder, *R. Buckminster Fuller: An Autobiographical Monologue Scenario Documented and Edited by Robert Snyder* [New York: St. Martin's, 1980], 12). On the effects of his time in the Navy, Fuller later explained: "You see how by this comprehensive anticipatory way of looking at things and thinking about the total needs of total man, I came a few years later to invent the words, 'Spaceship Earth.' Because I began to think about the total planet as being as beautifully designed and equipped as a ship. How do you run it in such a way as to take care of everybody?" (ibid., 29).

20. Fuller's work is full of computational metaphors. He often argued that the computer was an imitation of the human mind. See, e.g., R. Buckminster Fuller, *Operating Manual for Spaceship Earth* (Carbondale: Southern Illinois University Press, 1969), 112. In that same volume Fuller points to cybernetics and systems theory as key tools with which to solve the world's problems (87).

21. For a comprehensive and fascinating study of the role computers played in cold war psychology, and popular psychological discourse thereafter, see Paul N. Edwards, *The Closed World: Computers and the Politics of Discourse in Cold War America* (Cambridge, MA: MIT Press, 1996).

22. Fuller, quoted in Fuller and Snyder, *R. Buckminster Fuller*, 38.

23. Hugh Kenner, *Bucky: A Guided Tour of Buckminster Fuller* (New York: Morrow, 1973), 290.

24. Geof Bowker, "How to Be Universal: Some Cybernetic Strategies, 1943–1970," *Social Studies of Science* 23 (1993): 107–27; for a study of how these universal rhetorical strategies shaped the American counterculture see Fred Turner, *From Counterculture to Cyberculture: Stewart Brand, the Whole Earth Network, and the Rise of Digital Utopianism* (Chicago: University of Chicago Press, 2006).

25. For a historiography of the, relationship between the New Left and other movements of the time see Douglas Rossinow, "The New Left in the Counterculture: Hypotheses and Evidence," *Radical History Review* 67 (winter 1997): 79–120. See also Wini Breines, *Community and Organization in the New Left, 1962–1968: The Great Refusal* (New York: Praeger, 1982); and Douglas C. Rossinow, *The Politics of Authenticity: Liberalism, Christianity, and the New Left in America* (New York: Columbia University Press, 1998). For a broad overview of the era see Todd Gitlin, *The Sixties: Years of Hope, Days of Rage* (Toronto: Bantam, 1987).

26. By New Communalists I mean all of those who saw the transformation of individual consciousness rather than agonistic politics as the key to forming new communities. At the time, many who held such beliefs, including the back-to-the-landers, saw themselves as members of a "counter culture." Yet, in recent years, these beliefs have become entangled with drug use, sexual freedom, and hippie fashion under the same term. New Leftists, hippies, and communards have slowly begun to blur in our collective memory. By identifying the New Communalists as a separate subgroup within the social movements of the time, I hope to sharpen that historical picture.

27. Theodore Roszak, *The Making of a Counter Culture: Reflections on the Technocratic Society and Its Youthful Opposition* (Garden City, NY: Doubleday, 1969), 208.

28. Ibid., 50.

29. Ibid., 240.

30. Hugh Gardner, *The Children of Prosperity: Thirteen Modern American Communes* (New York: St. Martin's, 1978), 3. See also Mark Holloway, *Heavens on Earth: Utopian Communities in America, 1680–1880*, 2nd [rev.] ed. (New York: Dover, 1966).

31. Gardner, *The Children of Prosperity*, 9. Gardner also cites a *New York Times* survey that identified approximately two thousand "permanent communal living arrangements of significant size" across thirty-four states in 1970. Gardner argues that this figure should be seen as very conservative, since it does not include "small, urban cooperatives and collectives" (8). See also Richard Fairfield, *Communes U.S.A.* (San Francisco: Alternatives Foundation, 1971); William Hedgepeth, *The Alternative: Communal Life in New America* (New York: Macmillan, 1970); Robert Houriet, *Getting Back Together* (New York: Coward McCann and Geoghegan, 1971); Rosabeth Kanter, *Commitment and Community* (Cambridge, MA: Harvard University Press, 1972); Rosabeth Kanter, ed., *Communes: Creating and Managing the Collective Life* (New York: Harper and Row, 1973); Peter Rabbit, *Drop City* (New York: Olympia Press, 1971).

32. For a concise history of Drop City see Gardner, *The Children of Prosperity*, 35–48.

33. Quoted in ibid., 42. In a strange countercultural echo of the American government's cold war policy of dispersion, this gathering of domes was to be a prototype of an emerging world: "Soon domed cities will spread across the world, anywhere land is cheap—on the deserts, in the swamps, on mountains, tundras, ice caps. The tribes are

moving, building completely free and open way stations, each a warm and beautiful conscious environment. We are winning" (Rabbit, quoted in ibid., 37).

34. Rabbit, *Drop City*, 31; quoted in Hedgepeth, *The Alternative*, 36.

35. Alex Soojung-Kim Pang, "Dome Days: Buckminster Fuller in the Cold War," in *Cultural Babbage: Technology, Time and Invention*, ed. Francis Spufford and Jenny Uglow (Boston: Faber and Faber, 1996), 167–92.

36. Peggy [no surname], quoted in Hedgepeth, *The Alternative*, 153. For an incisive architectural analysis of the domes at Drop City see Simon Sadler, "Drop City Revisited," *Journal of Architectural Education* 59, no. 3 (2006): 5–14.

37. Stewart Brand, "Buckminster Fuller," in *The Last Whole Earth Catalog*, ed. Stewart Brand (Menlo Park, CA: Portola Institute, 1971), 3.

38. Gardner, *The Children of Prosperity*, 46–47.

Chapter 10

Research for this chapter was supported by a Research and Travel Grant, and a Humanities Center Grant, from the School of Humanities at the University of California, Irvine. I would also like to thank Hsiao-Yun Chu and the staff of Special Collections at Stanford University's Green Library for their kind assistance with the R. Buckminster Fuller Collection.

1. Gene Youngblood, "Earth Nova," *Los Angeles Free Press*, April 3, 1970, 34. Binelli was described here as a "young Israeli architect who had worked with Bucky in Ghana, teaching natives to build geodesic domes. He's now conducting a similar project with a group of street gangs in New York City."

2. See R. Buckminster Fuller, "Prevailing Conditions in the Arts," in *Utopia or Oblivion: The Prospects for Humanity* (New York: Bantam, 1969), 83–84.

3. R. Buckminster Fuller, "How It Came About (World Game)" (1969), repr. in R. Buckminster Fuller, *Your Private Sky: R. Buckminster Fuller, the Art of Design Science*, ed. Joachim Krausse and Claude Lichtenstein (Baden: Lars Müller, 2001), 472; hereafter *Your Private Sky: R. Buckminster Fuller, the Art of Design Science* is referenced as vol. 1 of *YPS*. "The world's increasing confidence in electronic instrumentation," he noted in another context, was "due to the demonstrated reliability of its gyrocompasses, and its 'blind' instrument landings of airplanes at night and in thick fog" (R. Buckminster Fuller, "The World Game: How to Make the World Work," in *Utopia or Oblivion: The Prospects for Humanity* [New York: Bantam, 1969], 160).

4. In a remarkable essay, Mark Wigley has traced Fuller's World Game project to military sources. He notes, for instance, that in 1941 "Fuller was part of [a] secret team of artists, filmmakers, designers and architects (including John Ford, Raymond Loewy, Walter Teague, Henry Dreyfuss, Norman Bel Geddes, Louis Kahn, Bertrand Goldberg, Lewis Mumford, and Walt Disney) working for the Visual Presentation Branch of the newly formed Office of Strategic Services (OSS, the predecessor of the CIA) to design a Presidential Situation Room that coordinates and efficiently presents 'a panorama of concentrated information' during war" (Mark Wigley, "Planetary Homeboy," *ANY* magazine, no. 17 [Jan. 1997]: 16–23).

5. Fuller, "How It Came About," 473.

6. Ibid.

7. Fuller, "The World Game," 161.

8. Fuller also attributed his popularization in the 1960s to Marshall McLuhan. See Thomas Albright, "Genius Who Kept His Optimism," *San Francisco Chronicle*, Nov. 23, 1967, 50, in R. Buckminster Fuller Collection, M1090:3:14, Stanford University Libraries (hereafter cited as Fuller MSS).

9. For an extended discussion of this topic see Felicity D. Scott, *Architecture or Techno-Utopia: Politics After Modernism* (Cambridge, MA: MIT Press, 2007).

10. R. Buckminster Fuller, "What Quality of Environment Do We Want?" *Archives of Environmental Health* 16 (May 1968): 699.

11. Fuller's presence can also be traced throughout the pages of underground papers such as the *San Francisco Oracle* and *East Village Other*.

12. "Evidently, Buckminster Fuller thinks design and planning would resolve the problems that politics has left unsolved for centuries," Maldonado noted, clarifying: "In other words, design and planning would be called in to substitute for politics, to abolish it and cancel it from history. 'Politics,' he says, 'will become obsolete.' It is not surprising, then, that he considers the 'Revolution by Design' to be exclusively an act of technical imagination: a position typical of technocratic utopianism" (Tomás Maldonado, *Design, Nature and Revolution: Toward a Critical Ecology*, trans. Mario Domandi [New York: Harper and Row, 1972], 29).

13. Gene Youngblood, "World Game: Escape Velocity," *Los Angeles Free Press*, May 15, 1970, 24.

14. The videosphere, he posited, "makes visible all the thoughts of all humanity simultaneously all around the planet. Not that TV is doing this now, but I think all of us can see that it could" (Gene Youngblood, "World Game: Scenario for World Revolution," *Los Angeles Free Press*, May 1, 1970, 42).

15. Youngblood, "World Game: Escape Velocity," 24.

16. Youngblood, "Earth Nova," 34. Fuller had been embraced in Britain earlier in the 1960s by architects like Archigram and the critic Reyner Banham. That this was Fuller's first encounter with "anti-Fuller propaganda" seems, however, rather unlikely.

17. Reproduced in *The Inflatable Moment: Pneumatics and Protest in '68*, ed. Marc Dessauce (New York: Princeton Architectural Press, 1999).

18. Posters depicting Fuller with a geodesic brain were at once an allegory of the militaristic ideology informing his geodesic domes and a parody of his anthropomorphizing claims. "It is no aesthetic accident that Nature enclosed our brains and regenerative organics in compoundly curvilinear structures," he earlier noted; "there are no cubical heads, eggs, nuts, or planets" (R. Buckminster Fuller, "The Age of the Dome," *Build International* 2, no. 6 [July–Aug. 1968]: 14).

19. Fuller, cited in Gene Youngblood, "World Game Part Two: The Ecological Revolution," *Los Angeles Free Press*, April 10, 1970, 55.

20. R. Buckminster Fuller, "Geosocial Revolution," in *Utopia or Oblivion: The Prospects for Humanity* (New York: Bantam, 1969), 177.

21. See Gene Youngblood, *Expanded Cinema* (New York: E. P. Dutton, 1970).

22. Youngblood, "World Game Part Two," 55.

23. Ibid.

24. Ibid.

25. Ibid.

26. Ibid., 58.

27. Ibid.

28. R. Buckminster Fuller, "How to Maintain Man as a Success in the Universe," in *Utopia or Oblivion: The Prospects for Humanity* (New York: Bantam, 1969), 248.

29. Youngblood, "World Game," 35.

30. Youngblood, "World Game: Escape Velocity," 24.

31. Fuller, "How It Came About," 473.

32. Fuller, "The World Game: How to Make the World Work," 157.

33. See Carl Schmitt, *The Concept of the Political*, trans. George Schwab (Chicago: University of Chicago Press, 1996).

34. Ibid., 158.

35. Carl Von Clausewitz, *On War* (1832), trans. Michael Howard and Peter Paret (Princeton, NJ: Princeton University Press, 1976).

36. See, e.g., Albright, "Genius Who Kept His Optimism," 50.

37. Fuller, "How It Came About," 479.

38. Fuller, "The World Game: How to Make the World Work," 159.

39. R. Buckminster Fuller, "Spaceship Earth—Fuller's World View," *Daily Californian*, Feb. 25, 1969, 9, Fuller MSS, M1090:3:15.

40. See Giorgio Agamben, *State of Exception*, trans. Kevin Attell (Chicago: University of Chicago Press, 2005).

41. "Clausewitz Updated," repr. in *The Movement Toward a New America*, ed. Mitchell Goodman (New York: Alfred A. Knopf, 1970), 578.

42. R. Buckminster Fuller, "Invisible Future," *San Francisco Oracle*, vol. 11, 1967, 21, Fuller MSS, M1090:2.

43. Fuller, *Utopia or Oblivion*, 269.

44. I am thinking here of the thesis of Jacques Rancière:

> The essence of politics resides in the modes of dissensual subjectification that reveal the difference of a society to itself. The essence of consensus is not peaceful discussion and reasonable agreement as opposed to conflict or violence. Its essence is the annulment of dissensus as the separation of the sensible from itself, the annulment of surplus subjects, the reduction of the people to the sum of the parts of the social body, and of the political community to the relationship of interests and aspirations of these different parts. Consensus is the reduction of politics to the police. In other words, it is the "end of politics" and not the accomplishment of its ends but, simply, the return of the "normal" state of things which is that of politics' non-existence. (Jacques Rancière, "Ten Theses on Politics," *Theory and Event* 5, no. 3 [2001]: http://muse.jhu.edu/journals/theory_and_event/v005/5.3ranciere.html [accessed Dec. 20, 2008])

45. Schmitt, *The Concept of the Political*, 55.

46. Ibid., 54.

47. See Branden W. Joseph, "Hitchhiker in an Omni-Directional Transport: The Spatial Politics of John Cage and Buckminster Fuller," *ANY* magazine, no. 17 (Jan. 1997): 40–44.

48. See Herbert Marcuse, *One-Dimensional Man: Studies in the Ideology of Advanced Industrial Society* (Boston: Beacon, 1964).

49. See Michael Hardt and Antonio Negri, *Empire* (Cambridge, MA: Harvard University Press, 2000).

50. Michael Hardt and Antonio Negri, *Multitude: War and Democracy in the Age of Empire* (New York: Penguin, 2004), 3.

51. Gene Youngblood, "World Game," *Los Angeles Free Press*, May 29, 1970, 34.

52. Peter L. Douthit, "Drop City: A Report from the Energy Center," *Arts Magazine* 41 (1967): 233.

53. Letter from Drop City to R. Buckminster Fuller, Jan. 12, 1967, in Fuller MSS, M1090:2:141. The letter was signed Peter L. Douthit, Burt Wachman, Gene Bernofsky, John Fudge, Joe Clower, John Ceul, Judy Douthit, Charles DiJulio, Carol DiJulio, JoAnn Bernofsky, Clark Richert, Richard Xallweit, and Jill Speed.

54. John McHale to Peter L. Douthit, Nov. 9, 1966, Fuller MSS, M1090:2:145.

55. Douthit, "Drop City," 49.

56. Albin Wagner, "Drop City," *Avatar*, Aug. 4–18, 1967, 7.

57. Douthit, "Drop City," 50.

58. Wagner, "Drop City," 7.

59. Peter Rabbit, *Drop City* (New York: Olympia Press, 1971), 28–29.

60. Baer's *Dome Cookbook* was published by the Lama Foundation, a not-for-profit dome-building community in New Mexico, which would also later reprint Fuller's *4D Time Lock* of 1927. Baer's manual was dedicated to the use of convex figures, in particular the family of polyhedrons known as zonahedra, cast as a contribution to architecture's future outside the extant "military-industrial-educational complex." Along with a main text addressing the geometric qualities of various polyhedra and their ability to interlock and close-pack, *Dome Cookbook* included handwritten notes offering feedback and secondary observations. A number were in critical, and somewhat paranoid, dialogue with Fuller. "Today you hear people who are said to be ahead of their time say 'the world is a huge space ship,'" he wrote. The problem was that spaceships "are built by men but this planet was not." To Baer such rhetoric simply represented an urge "to take over and completely control every variable of our environment—to in fact reform our planet into a giant space ship—living machine" (Steve Baer, *Dome Cookbook* [Coralles, NM: Lama Foundation/Cookbook Fund, 1968], 8).

61. Lloyd Kahn Family to R. Buckminster Fuller, April 9, 1969, Fuller MSS, M1090:2:184. Fuller was in the Bay Area in November 1967 for a conference titled "Designing Environments for Expanding Awareness," sponsored by the Esalen Institute at Big Sur. See Albright, "Genius Who Kept His Optimism," 50.

62. *Domebook 1* (Los Gatos, CA: Pacific Domes, 1970), 3.

63. All citations in this paragraph are from Kahn's letter of April 9, 1969.

64. Dale Klaus to Lloyd Kahn, June 8, 1970, Fuller MSS, M1090:2:203.

65. Dale Klaus to Jay and Kathleen Baldwin, June 8, 1970, Fuller MSS, M1090:2:203.

66. Lloyd Kahn, ed., *Domebook 2* (Bolinas, CA: Pacific Domes, 1971), 65.

67. Ibid., 64–65.

68. Ibid.

69. "The Wonder of Jena," in *Shelter*, ed. Lloyd Kahn (Bolinas, CA: Shelter Publications, 1973), 111.

70. Lloyd Kahn, "Smart but Not Wise," in *Shelter*, ed. Lloyd Kahn (Bolinas, CA: Shelter Publications, 1973), 113.

71. Youngblood, "World Game: Escape Velocity," 24.

72. Youngblood, "World Game Part Two," 55.

73. Gene Youngblood, "Bucky in the Universe," *Los Angeles Free Press*, June 5, 1970, 33. In this short article Youngblood stressed the importance of the collected essays published as *Utopia or Oblivion*. He also recounted a conversation with Fuller, noting that "I had never in my life felt so apocalyptic." If *Utopia or Oblivion* served as a motivating force for this embrace of evolution through technology, *Operating Manual for Spaceship Earth*, also published in 1969, would serve as the movement's manifesto. See R. Buckminster Fuller, *Operating Manual for Spaceship Earth* (Carbondale: Southern Illinois University Press, 1969).

74. Fuller, "Age of the Dome," 7.

Chapter 11

1. Walter Benjamin, "Theses on the Philosophy of History" (1939), in *Illuminations*, ed. Hannah Arendt, trans. Harry Zohn (New York: Schocken Books, 1969), 253–64.

2. Although he attempted to position himself outside what he took to be the ideological impasse of the cold war, Fuller worked on a number of projects for the U.S. military during the 1950s and 1960s. Among these were the "radomes"—geodesic domes designed by Fuller to house radar installations along the Defense Early Warning (DEW) Line near the Arctic Circle, which were part of the Semi-Automated Ground Environment (SAGE) continental air defense system.

3. See Daniel Bell, *The End of Ideology: On the Exhaustion of Political Ideas in the Fifties* (Glencoe, IL: Free press, 1960); and Daniel Bell, *The Coming of Post-industrial Society: A Venture in Social Forecasting* (New York: Basic Books, 1973).

4. Charles Jencks, *The Language of Postmodern Architecture* (London: Academy Editions, 1977), 9.

5. Charles Jencks, *Late-Modern Architecture and Other Essays* (New York: Rizzoli, 1980), 65.

6. Fredric Jameson, *Postmodernism, or, The Cultural Logic of Late Capitalism* (Durham, NC: Duke University Press, 1991), 305.

7. Fredric Jameson, *A Singular Modernity: Essay on the Ontology of the Present* (London: Verso, 2002), 165–73.

8. Jameson, *Postmodernism*, 51–52, 415–17.

9. Fredric Jameson, "The Politics of Utopia," *New Left Review* 25 (Jan.-Feb. 2004): 50–51.

10. Jean-François Lyotard, *The Postmodern Condition: A Report on Knowledge*, trans. Geoff Bennington and Brian Massumi (Minneapolis: University of Minnesota Press, 1984), 9–10.

11. Ibid., 59.

12. See Reinhold Martin, "Crystal Balls," *ANY* magazine, no. 17 (Jan. 1997): 35–39.

13. Ibid., 66.

14. R. Buckminster Fuller, "World Game: How It Came About," in *50 Years of the*

Design Science Revolution and the World Game (Carbondale: World Resources Inventory Southern Illinois University, 1969), 112.

15. Ibid., 114.

16. Ibid., 111.

17. Ulrich Beck, *Risk Society: Towards a New Modernity*, trans. Mark Ritter (London: Sage, 1992), 21.

18. Ibid., 34.

19. R. Buckminster Fuller, *Operating Manual for Spaceship Earth* (New York: Simon and Schuster, 1969), 58–59.

20. *Just Gaming* was the title given to a series of interviews between Lyotard and Jean-Loup Thébaud on the subject of language games and justice. See Jean-François Lyotard and Jean-Loup Thébaud, *Just Gaming*, trans. Wlad Godzich, afterword by Samuel Weber (Minneapolis: University of Minnesota Press, 1985). In Lyotard's terms to the degree that Fuller's World Game would qualify as a language game, it would be an "impure" one in that it mixes descriptive / denotative and prescriptive statements, or technical statements made by "experts" and statements that prescribe a "just" (or in Fuller's terms, a utopian) outcome based on these technical descriptions.

21. Jameson, *Postmodernism*, 40.

22. Ibid., 44.

Index

Italicized page numbers refer to material in illustrations.